Logic and Information Flow

Foundations of Computing

Michael Garey and Albert Meyer, editors

Complexity Issues in VLSI: Optimal Layouts for the Shuffle-Exchange Graph and Other Networks, Frank Thomson Leighton, 1983

Equational Logic as a Programming Language, Michael J. O'Donnell, 1985

General Theory of Deductive Systems and Its Applications, S. Yu Maslov, 1987

Resource Allocation Problems: Algorithmic Approaches, Toshihide Ibaraki and Naoki Katoh, 1988

Algebraic Theory of Processes, Matthew Hennessy, 1988

PX: A Computational Logic, Susumu Hayashi and Hiroshi Nakano, 1989

The Stable Marriage Problem: Structure and Algorithms, Dan Gusfield and Robert Irving, 1989

Realistic Compiler Generation, Peter Lee, 1989

Single-Layer Wire Routing and Compaction, F. Miller Maley, 1990

Basic Category Theory for Computer Scientists, Benjamin C. Pierce, 1991

Categories, Types, and Structures: An Introduction to Category Theory for the Working Computer Scientist, Andrea Asperti and Giuseppe Longo, 1991

Semantics of Programming Languages: Structures and Techniques, Carl A. Gunter, 1992

The Formal Semantics of Programming Languages: An Introduction, Glynn Winskel, 1993

Exploring Interior-Point Linear Programming: Algorithms and Software, Ami Arbel, 1993

Hilbert's Tenth Problem, Yuri V. Matiyasevich, 1993

Theoretical Aspects of Object-Oriented Programming: Types, Semantics, and Language Design, edited by Carl A. Gunter and John C. Mitchell, 1993

From Logic to Logic Programming, Kees Doets, 1994

The Structure of Typed Programming Languages, David A. Schmidt, 1994

Logic and Information Flow, edited by Jan van Eijck and Albert Visser, 1994

Logic and Information Flow

edited by
Jan van Eijck and Albert Visser

The MIT Press
Cambridge, Massachusetts
London, England

© 1994 Massachusetts Institute of Technology

This book was set in LaTeX by the authors and was printed and bound in the United States of America.

Library of Congress Cataloging-in-Publication Data

Logic and information flow / edited by Jan van Eijck and Albert Visser.
    p.   cm. — (Foundations of computing)
  Includes bibliographical references.
  ISBN 0-262-22047-4
  1. Computer science. 2. Natural language processing (Computer science) 3. Logic, Symbolic
and mathematical. I. Eijck, J. van (Jan) II. Visser, Albert. III. Series.
QA76.L5662   1994
511.3—dc20   93-40778
                                                    CIP

Contents

Series Foreword

Theoretical computer science has now undergone several decades of development. The "classical" topics of automata theory, formal languages, and computational complexity have become firmly established, and their importance to other theoretical work and to practice is widely recognized. Stimulated by technological advances, theoreticians have been rapidly expanding the areas under study, and the time delay between theoretical progress and its practical impact has been decreasing dramatically. Much publicity has been given recently to breakthroughs in cryptography and linear programming, and steady progress is being made on programming language semantics, computational geometry, and efficient data structures. Newer, more speculative, areas of study include relational databases, VLSI theory, and parallel and distributed computation. As this list of topics continues expanding, it is becoming more and more difficult to stay abreast of the progress that is being made and increasingly important that the most significant work be distilled and communicated in a manner that will facilitate further research and application of this work. By publishing comprehensive books and specialized monographs on the theoretical aspects of computer science, the series on Foundations of Computing provides a forum in which important research topics can be presented in their entirety and placed in perspective for researchers, students, and practitioners alike.

Michael R. Garey
Albert R. Meyer

Preface

This is a book on as many possible aspects of the dynamics of information flow as we could think of, from a logical point of view. The logic of information flow has applications in computer science and in natural language analysis, but it is also a budding branch of logic *per se* ('logic of action'), and finally there are philosophical ramifications. All of these topics are explored at some length, with a bit more emphasis on logical techniques and on applications than on philosophy. A very interesting feature of almost all of the chapters is that they are trying to bridge gaps, chart landscapes or draw maps. The introductory chapter gives more information on how the individual chapters are connected.

This book was written with three groups of readers in mind. It is meant for philosophers with an interest in the nature of information and action, and in the relation between those concepts. It is aimed at theoretical computer scientists with an interest in up-to-date formalisms of 'dynamic logic', and their possible applications. Finally, it is relevant for logicians with an interest in a broadening of their discipline beyond the realm of 'sound reasoning' in the narrow sense.

How can logic, the study of reasoning, be applied to information flow? Information, in its most general sense, has something to do with being adapted to an environment. An organism or object O is adapted to its environment E if changes in the environment are somehow reflected in the organism or object. Next, if a second object O' is adapted to changes in the first object O, then we can say that there is an indirect information link between E and O. Thus, logic can study information in at least two different ways: by studying the nature of the adaptation, and by studying the nature of the linking. Also the logical study of information can occur in different settings, such as theoretical computer science, philosophy, or natural language interpretation. Finally, the logical analysis can make use of a variety of logical tools, such as algebra, modal logics, or various brands of dynamic logic.

The book chapters were written expressly for this volume by logicians, theoretical computer scientists, philosophers and semanticists, to explore all these variations and chart some of the connections. The starting point for this enterprise was a workshop on Logic and the Flow of Information, which was held at 'Hotel de Filosoof' in Amsterdam, in December 1991.

We wish to acknowledge a grant from the Netherlands Organization for for the Advancement of Research (N.W.O.), who sponsored the workshop under project NF 102/62-356 ('Structural and Semantic Parallels in Natural Languages and Programming Languages'). Also, we would like to thank Robert Prior from MIT Press for his support and enthusiasm during the final preparation stages of this book.

Logic and Information Flow

1 Logic and Information Flow

Jan van Eijck and Albert Visser

1.1 Kinds of Information Flow

The volume that you are holding now, dear reader, may affect your state of knowledge concerning the topic of *Logic and Information Flow* in at least three different ways. First, assuming you are reading this well before the turn of the millennium, it will provide you with an up-to-date impression of different kinds of logics that all somehow address the topic of change and dynamics. Secondly, it will teach you differences. The chapters of this volume show that there are many different ways in which logic can broaden its horizon to a dynamic perspective, and that often several of those can be combined in one system. Finally, it will provide you with lots of hints concerning possible connections between all those different approaches to change, dynamics and information flow.

The following chapters report on a wide range of logics that are part of a wide class of traditions. Read as exponents of those traditions, they provide us with introductions to process algebra, to a variety of action logics, to generalizations of and variations on propositional dynamic logic, to an epistemic perspective on logic with a link to connectionism, to the relation between dynamic predicate logic and ordinary predicate logic, to attempts at modelling dynamic aspects of natural language semantics, such as pronoun–antecedent links and presuppositions, and finally, to the abstract study of the way in which information may function as a link between different kinds of situations.

It might be helpful if we draw two distinctions to assist the reader in establishing links between the contributions for her or himself, and to save her or him from unnecessary conceptual confusion. Two kinds of information flow play a role in this book:

- information flow within one site: as a result of reasoning (the province of logic in the narrow sense) or absorbing new facts (also within the province of logic, in a broader sense).

- information flow from one side to another: bringing to bear facts and rules about one situation in another situation (the province of cognitive science, or of logic in a still broader sense).

Another important distinction is that between two kinds of dynamics that are subject of investigation in the pages that follow.

- the dynamics of information change, i.e., the study the process of information increase or information loss by means of acts of absorbing new information, belief revision on the basis of new facts, etcetera.

- the logical study of systems involving change, such as processes (this subsumes the previous item, for the process of information change can be viewed as such a system involving change, but it is much wider).

All authors have taken great trouble to relate their contribution to other work, but they do so, of course, from their own perspective. One contribution (Pratt's *Roadmap*) was even explicitly written to trace a number of important relations between action logics.

1.2   Themes, Approaches, Topics

We will not attempt to relate every chapter to every other chapter in this volume. Instead, we will trace some of the links we see, while inviting the reader to reflect on further completions of the pattern. We distinguish the following themes in this volume:

- Dynamic reconstructions of well known logics. (Gärdenfors)
- Uses of logics in information processing. (Moss and Seligman)
- Reasoning about change and process behaviour. (Kozen, Pratt, Ponse)
- Development of logics with both static and dynamic modes: enrichments of propositional dynamic logic. (De Rijke)
- Development of logics with both static and dynamic modes, but based on a poorer set of assumptions: dynamic arrow logic. (Van Benthem)
- Modelling dynamic aspects of natural language semantics.
(Van Eijck, Kracht, Visser)

We can also make a rough distinction in approaches or traditions, as follows.

- Relational algebra approach. (Kozen, Pratt, Visser, Andreka c.s.)
- Process algebra approach. (Ponse)
- Situation theory approach. (Moss and Seligman)
- Update logic approach. (Gärdenfors, Visser)
- Modal logic / dynamic logic approach. (Van Benthem, Van Eijck, De Rijke)
- Partial logic approach. (Kracht)

Still another dimension is the main topic or emphasis of the chapters. If we measure them along this dimension, a third partition emerges, which looks somewhat like this.

- New logics for dynamic processes. (Kozen, Van Benthem)
- Philosophical claims about existing logics. (Gärdenfors)
- Connections between logics or approaches. (Van Eijck, Moss and Seligman, Pratt, De Rijke, Ponse)
- Analysis of presuppositions. (Kracht, Visser)

The fact that half of the chapters have comparison or classification as their main topic is of course a reflection of the theme of the workshop that gave rise to these contributions.

1.3   An Overview of the Individual Chapters

We now give some short introductory comments on each of the chapters, with pointers to other chapters in the volume wherever appropriate.

Van Benthem   Arrow logic is a kind of dynamic logic or action logic based on a weaker set of underlying assumptions than usual. In propositional dynamic logic or relational algebra, every program relation brings a converse relation in its wake, and every pair of relations a composition. The idea is to get rid of these facts by dropping the analysis of transitions as composed of pairs of states. Taking the transitions themselves as primitive and calling them arrows has the advantage that we can stipulate what relations hold between them: composition and reversal themselves become two place relations on the set of arrows. Similarly, identity becomes a property, holding of those arrows that end at the place where they started.

To understand what is going on, it is illuminating to compare the notion of similarity between structures that underlies propositional dynamic logic with the similarity notion of arrow logic. In propositional dynamic logic, state s in structure M is bisimilar to state r in structure N if s and r have the same propositional valuation, and every atomic program step a from s to a state s' can be matched by an a step from r to some r' with the property that s' and r' are bisimilar, plus the same clause vice versa.

In arrow logic, bisimilarity becomes a relation between arrows. Arrow a in structure M is bisimilar to arrow b in structure N if a and b have the same valuation, and the following hold:

●

– If there are arrows a_1, a_2 in M such that a_2 is a composition of a and a_1 (a_2 is a composition of a_1 and a, a is a composition of a_1 and a_2) then there are arrows b_1, b_2 in N such that b_2 is the composition of b and b_1 (b_2 is a composition of b_1 and b, b is a composition of b_1 and b_2), and a_1 is bisimilar to b_1, a_2 bisimilar to b_2.

– If there is an arrow a_1 in M such that a_1 is a reversal of a, then there is an arrow b_1 in N such that b_1 is a reversal of b, and a_1 and b_1 are bisimilar.

– If a is an identity arrow, then b is an identity arrow as well.

● Same clause vice versa.

The connection with action algebra is pointed out by Van Benthem himself in one of the appendices of the chapter. First, the two residuation operators familiar from categorial logic and from Pratt's action logic, are definable in basic arrow logic (arrow logic with composition and reversal, but without Kleene star). $a\backslash b$ (in Pratt's notation: $a \rightarrow b$) is short for $\neg(a\check{\,} \cdot \neg b)$. If we write this out, the interpretation for $a\backslash b$ becomes: $M, x \models a\backslash b$ if for all y, z decomposing x it holds that $M, r(y) \models a$ implies $M, z \models b$

(where $r(y)$ denotes the reversal of the y arrow). Second, the principle $(a\backslash a)^* = (a\backslash a)$ is derivable in arrow logic.

Of course, Pratt introduces the two residuation operators in order to get right and left inverses for composition without having to commit himself to a dangerous negation operator (dangerous, because it would threaten the decidability of the logic). Arrow logic does have negation, and still it is decidable. How come? Briefly, the reason is that in arrow logic negation has lost most of its sting. $\neg(\varphi \circ \psi)$ is true of all arrows x that cannot be decomposed in two arrows y, z with y satisfying φ and z satisfying ψ. In relation algebra, $\neg(\varphi \circ \psi)$ means something rather different (and much stronger): it is true of all pairs of states $\langle s_1, s_2 \rangle$ *except* for those for which there is a state s_3 with $\langle s_1, s_3 \rangle$ satisfying φ and $\langle s_3, s_2 \rangle$ satisfying ψ.

Van Eijck Van Eijck starts out with a problem from natural language semantics: pronouns that are linked to their antecedents without being bound by those antecedents in the traditional logical sense. He investigates a framework that was proposed to deal with such cases: a version of predicate logic with dynamic binding principles, so called dynamic predicate logic.

It turns out that a version of quantified modal logic with the formulae of dynamic predicate logic as programs can be used to analyse the connection between dynamic predicate logic and ordinary first order logic. What emerges is a perspective on the semantics of natural language, where natural language meanings are encoded in a dynamic representation language which has all the 'hooks' in place for linking pronouns to their antecedents. At any given point in the processing of discourse, one might move from these dynamic representations to static truth conditions, using essentially a form of precondition reasoning on the dynamic representations.

Van Eijck shows that quantified dynamic logic over dynamic predicate logic is only two dimensional (to adopt the terminology of Pratt's contribution to this volume) in appearance. Every formula involving DPL modalities is equivalent to a first order formula, and the axiomatisation given in the chapter provides a straightforward algorithm to remove the DPL modalities.

Dynamic predicate logic was introduced to deal with one of the simplest dynamic aspects of natural language semantics: pronoun antecedent links. Other dynamic aspects are: presuppositions of natural language constructs and their dependence on a dynamically changing context, and the study of the way in which natural language utterances change the state of knowledge or belief of the addressee, as they are being interpreted and absorbed within a given context that is itself changed by the incoming utterances. The chapters of Kracht and Visser address aspects of the problem of presuppositions. The contribution of Visser also touches upon the wider issue of the information change potential of utterances.

Gärdenfors   Gärdenfors starts by making a distinction between a static view of logic and an update view of logic. His thesis is that the update view of logic, where meanings are defined in terms of their information change potential, does not leave out anything. This thesis is illustrated by a demonstration how (intuitionistic or classical) propositional logic emerges from a set of update postulates. Gärdenfors then moves on to make a further distinction, between a symbolic approach and a connectionist approach to information processing. Here his thesis is that these approaches can be reconciled. This is illustrated by the development of a logic of schemata within a connectionist framework.

If we were to set topics for term papers on this subject, we would suggest the following further questions. How about developing the update view of logic beyond the propositional connectives? What would operations for expansion or contraction of information look like? Are there still static counterparts for these? A static counterpart for 'expand with φ' could be a pair of descriptions of the states before and after the expansion operator. What is the appropriate language for this? There is a link here to questions posed in the chapter of De Rijke. The same questions could be posed for the schemata logic that emerges from the analysis of connectionist networks.

Kozen   Kozen investigates the relation between Kleene algebras and action algebras. Kleene algebras have the nice property that they are closed under the formation of matrices. Matrices are very useful for making the link between Kleene algebras and automata. Consider the automaton of Figure 1.1.

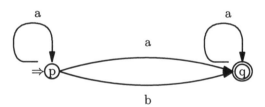

Figure 1.1
Finite state automaton

The transitions for this automaton can be given in the form of a matrix, where position $(0,0)$ in the matrix gives the transitions from p to p, $(0,1)$ the transitions from p to q, $(1,0)$ those from q to p, and $(1,1)$ those from q to q.

$$\begin{bmatrix} a & a+b \\ 0 & a \end{bmatrix}$$

In this transition matrix, the rows indicate where the transitions start, the columns where they end. We can point at start states and accepting states by means of $0, 1$ single row or single column matrices.

If we want to know which transitions are possible from the start state we simply perform matrix multiplication:

$$[1\ 0] \cdot \begin{bmatrix} a & a+b \\ 0 & a \end{bmatrix} = [a\ a+b].$$

Similarly if we want to know which transitions end in an accepting state:

$$\begin{bmatrix} a & a+b \\ 0 & a \end{bmatrix} \cdot \begin{bmatrix} 0 \\ 1 \end{bmatrix} = \begin{bmatrix} a+b \\ a \end{bmatrix}$$

Now how do we get the reflexive transitive closure of the relation given by the transition matrix? For that Kozen defines the Kleene star operation on matrices, as follows. Assume

$$E = \begin{bmatrix} a & b \\ c & d \end{bmatrix}$$

and $f = a + bd^*c$. Then $E*$ is given by:

$$E^* = \begin{bmatrix} f^* & f^*bd^* \\ d^*cf^* & d^* + d^*cf^*bd^* \end{bmatrix}.$$

To understand this, think of the two state automaton for which E is the transition matrix. For note that a gives the transitions from p to p, b those from p to q, c those from q to p, and d those from q to q. Now $f^* = (a + bd^*c)^*$ gives you all the paths from p to p, for if you start out from p you have a choice between either performing the a loop or going to q, looping there for a while, and coming back again. Similarly, f^*bd^* gives you all the paths from p to q. For any such path consists of first pursuing a path beginning at p and ending at p, then moving from p to q, and finally looping at q. Note again that this is indeed an exhaustive description. All the paths from q to p are given by d^*cf^*, and finally, all the paths from q to q are given by $d^* + d^*cf^*bd^*$.

This definition is next generalized inductively to arbitrary $n \times n$ matrices, by partitioning a matrix E into submatrices A, B, C and D, with A and D square.

$$E = \left[\begin{array}{c|c} A & B \\ \hline C & D \end{array} \right]$$

Using the fact that D^* has already been defined because it is square and of smaller size, F can be defined as $A + BD^*C$, and since F has the same size as A, F^* has already been defined and can be used in the definition of E^*.

Using the definition of the Kleene star operation on matrices, we derive the language accepted by the machine from Figure 1.1 as follows:

$$[1\ 0] \cdot \begin{bmatrix} a & a+b \\ 0 & a \end{bmatrix}^* \cdot \begin{bmatrix} 0 \\ 1 \end{bmatrix} \ = \ [1\ 0] \cdot \begin{bmatrix} a^* & a^*(a+b)a^* \\ 0 & a^* \end{bmatrix} \cdot \begin{bmatrix} 0 \\ 1 \end{bmatrix}$$

$$= \ a^*(a+b)a^*.$$

Now the connection with action algebra is simple. We have matrix addition, matrix multiplication and Kleene star on matrices, so what we need is appropriate left and right residuation operations on square matrices. The main point of the chapter is that these can only be defined in an action algebra on condition that the algebra has finite meets (i.e., is a lattice). For then we can do it componentwise, as follows ($\prod$ is used for iterated, but finite, meet):

$$(A \to B)_{ij} \ = \ \prod_{k=1}^{n} (A_{ki} \to B_{kj}).$$

And similarly for left residuation.

Kracht Kracht studies the information flow of presuppositions. The chapter starts with a discussion of example sentences from natural language and from informal mathematical discourse with the peculiar property that certain statements can be derived from these sentences and from their negations. *The present king of France is not bald* entails that France has a king, but so does the negation of this sentence (if we assume that both sentences are uttered with normal stress).

Sentences such as *The present king of France is not bald* only express propositions on condition that their presupposition (in this case: *France has a king*) are fulfilled. There is a vast linguistic literature on presupposition to which Kracht refers. He is not completely happy, however, with current logical accounts of presupposition.

Kracht presents a fresh attempt at incorporating presupposition into logic, after some careful distinctions that are not often made in the literature, e.g., between the projection problem and the separation problem. The separation problem is the problem of separating statements from their presuppositions in such a way that the presupposition itself does not have presuppositions of its own. In other words, an answer to the separation problem is a way of specifying presuppositions that themselves express propositions in all circumstances. The projection problem is the problem of deriving the presupposition of a statement from the assertions and presuppositions of its component statements.

The division of sentences in assertive and non-assertive (with presuppositions that can be fulfilled or not fulfilled) engenders a distinction in admissible and unwanted truth values. The truth values *true* and *false* can be used to make assertions, so they are admissible. All other truth values cannot be used to make assertions, so they give rise to presupposition failure.

Kracht attacks the problem in very general terms, by developing a theory of information flow in logical operators by means of 'information locks' and bringing the resulting logic to bear on the analysis of presuppositions. Suppose we work in a three valued logic (where the third value U is read as 'unknown'), and we want to extend the logical connectives to the three valued case. For a two place propositional connective $f(\_1, \_2)$ there are four possibilities. If the information lock between the two input positions is closed, we get the weak Kleene extension of the two valued truth table: a value U in either of the input positions will yield output value U. In case the information lock between the two input positions is open in both directions, we get the strong Kleene extension of the two valued truth table. Two inputs U will yield output U, but a definite value in one of the input slots may force a definite result. Finally, the lock can be open in the left-right or in the right-left direction. For instance, if the lock is open from left to right but not vice versa, the truth table for $\vee$ will have $T \vee U = T$ and $U \vee T = U$, in the other case it is the other way around. Information locks open in the left to right direction seem to be the staple of linguistic processing. They are also familiar to those acquainted with the lambda-K-calculus. The concept of information locks leads to a plethora of new logical connectives in three valued logic.

After these preliminaries, it is straightforward to define a two place connective which separates out presuppositions from assertions: $p \downarrow q$ yields the value of q in case p is true, the value U otherwise. This connective is essentially different from the propositional connectives derived from the two valued ones by setting the information lock, for these all have the property that definite values for the inputs yield a definite output value. Kracht in fact develops a theory of presuppositions for three valued logic by giving an algorithm for deriving presupposition normal forms for formulae of the three valued language, and showing how the 'most general' presupposition of a formula can be derived from this normal form.

For an example application to natural language, suppose we have a discourse such as: *The king of France is bald. He wears a wig.* Suppose we know that the presupposition of the first sentence is that France has a unique king. Then we can use Kracht's calculus to derive that this will also be the presupposition of the whole discourse.

To connect this theory of presupposition with Van Eijck's chapter on dynamic predicate logic, we briefly mention an extension of dynamic predicate logic with error states. Presupposition failure of a proposition A in a given context causes the generation of an error state ϵ. The semantic definition would run like this (for a statement A with presupposition P and assertion B):

$$[\![A]\!](s) = \begin{cases} \{s\} & \text{if } s \models P \text{ and } s \models B \\ \emptyset & \text{if } s \models P \text{ and } s \not\models B \\ \{\epsilon\} & \text{if } s \not\models P. \end{cases}$$

Then a calculus along the lines of Van Eijck's chapter may be devised yielding the 'most general' preconditions for truth and falsity of all formulas in which A occurs. Such a calculus would engender the following. $\langle A \rangle \top \leftrightarrow P \wedge B$, giving the preconditions for:

only proper outputs, plus at least one output. $[A]\bot \leftrightarrow P \wedge \neg B$, giving the preconditions for: no outputs at all. The preconditions for presupposition failure, only ϵ as output, would then be given by: $(\neg\langle A\rangle\top \wedge \neg[A]\bot) \leftrightarrow P$. Presupposition projection can be accomplished in this framework by applying rules for concatenation, dynamic quantifiers and other operators that yield their preconditions. As an additional bonus, the resulting theory also nicely accounts for the link between the pronoun *he* and its antecedent *the king of France* in the example discourse.

It would be most interesting to explore the connection between such an approach and Kracht's three valued logic of presuppositions.

Moss and Seligman   It is important to note that the kind of dynamics analysed here is timeless. It is the dynamics of making abstract links between one kind of situation and another, the dynamics of absorbing facts about one situation, and bringing them to bear in another one.

In this approach, logic is characterized as information flow at one site. This shows up in the Barwise rule of 'logic as information flow', which essentially expresses soundness.

The chapter is a survey of related approaches, pointing out the connections between those. It traces a connection between logic and topology, where a proposition is analysed as an open set. It gives a modal logic of topology, and gives some examples of its use. It discusses Barwise's rules of information linking, and hints at connections with category theory and ideas from constructivism.

Information is perceived as a means for classifying, by linking sites (classified objects) to types (classifying objects). There is a duality here: one might just as well look at sites as classifiers of types. Information linking is making connections between levels. Pixels on a screen are connected to patterns of characters. The character pattern is connected to a description in English. The description is linked to a snapshot picture. The snapshot picture is connected to the situation where it was taken. Lots of different links are possible at the same time. Dogshit on my doorstep is linked to my neighbour's dog, but also to a brand of dogfood, to my change of mood when I open the front door and take a deep breath, etcetera.

The chapter gives glimpses of a landscape that invites further exploration. One would like to know more about the properties of the modal logic of topology. Also, one would like to see what is behind the particular set of Barwise rules for information links. There are clear connections to well-known principles. For instance, the Xerox principle (that allows one to conclude from a link between pixels on a screen and a pattern of characters and a link between the character pattern and a message in English to a link between the pixel on the screen and the message) looks suspiciously like a cut rule. Does this mean that there also is a notion of cut elimination of information linking?

Perhaps the easiest way to see that information flow in the sense of Moss and Seligman is different from dynamics in the sense of dynamic logic is by considering a system of information links over some form of dynamic logic. Here the subject matter is processes (as they are studied by dynamic logic), but the linking is between processes in different situations. Barwise's Xerox principle now applies to the links between a flow chart of a coffee machine and the external behaviour of the machine on one hand and that between the external behaviour machine and its internal mechanics on the other. It allows to conclude from these links to a link between the flow chart and the internal mechanics of the coffee machine.

Ponse   Ponse's chapter contains a general introduction to Process Algebra in the Amsterdam tradition, and then extends the operations on processes with test operations which he calls guards.

Ponse comments on the differences between process algebra and dynamic logic, saying that there is a difference between equational reasoning with terms denoting processes (as in Process Algebra) and describe properties of processes by means of formulas of an appropriate language (as in propositional dynamic logic). He gives the example of the basic process algebra axiom expressing the commutativity of choice.

$$x + y = y + x.$$

Comparing this with propositional dynamic logic, there seems to be a natural connection with the following formula.

$$\langle x \cup y \rangle \varphi \leftrightarrow \langle y \cup x \rangle \varphi.$$

This is indeed valid in propositional dynamic logic. But the last word about the precise connection has not been said, for we in process algebra we have:

$$x(y + z) \neq xy + yz.$$

Compare this to propositional dynamic logic, where the following is a valid principle.

$$\langle x; (y \cup z) \rangle \varphi \leftrightarrow \langle x; y \cup x; z \rangle \varphi.$$

But such comparisons are only a side issue in Ponse's contribution. His main concern is to work out a system where process algebra is enriched with boolean tests, with the purpose of implementing precondition reasoning, as in Hoare logic. In fact, the system demonstrates by example how boolean testing can be incorporated within the framework of process algebra.

What Ponse proposes can be seen as an attempt to combine systems of reasoning at two different levels. At a global level, Ponse wants to handle transitions between states, and he does this in terms of process algebra operations and equalities. At a local level, Ponse want to reason about the internal structure of the states, by means of his logic of guards. This combination of two levels of reasoning within one system reminds one of a proposal by Finger and Gabbay (*Adding a Temporal Dimension to a Logic System*[1]) for a canonical way to amalgamate two logics L_1, L_2 into one and use the resulting logic to reason about a two-level structure where each of the logics described one of the levels. In the context of the chapter of Ponse, the logic that would be needed to study processes with guards seems to have the same two level structure:

- many sorted modal logic to reason about processes;
- first order logic to reason about what happens within states.

The challenge is to first find the right connection between process algebra and many sorted modal logic, then combine logics *à la* Finger and Gabbay, and finally describe the connection between this logic and Ponse's processes with guards. The two–layer logic that is the missing link has a first order logic part wrapped inside many sorted modal logic, in the obvious sense that no modal operators are allowed within the scope of quantifiers.

Pratt   Pratt starts by introducing the notion of a two dimensional logic, which is essentially a logic which talks about two independent orderings of a universe. Modal logic is an example as it talks about valuations on worlds (or, the boolean structure of worlds) on one hand, and the relational ordering between worlds on the other, and neither dimension is reducible to the other one.

As it turns out, two dimensional logic has a long history, which is traced back by the author to Aristotle, and extended all the way to Kozen's action lattices in the present volume, with due credits along the road given to De Morgan, Peirce, Schröder, and others.

Viewed from the perspective of full relation algebra (which Pratt calls starred relation algebra), the roadmap pictures all the different algebras one gets by considering reducts of this, i.e., by considering the algebra's for the various subsets of the vocabulary of relation algebra. The operations of starred relation algebra are: composition (denoted by ;), union, star (transitive closure under composition), residuation (two semi inverses for composition), complement (Boolean negation), and converse.

In relation algebra the two residuation operators are definable, for we can define left and right inverses for composition as follows:

$$a \backslash b \stackrel{\text{def}}{=} \neg(a^{\smile}; \neg b)$$

$$a/b \stackrel{\text{def}}{=} \neg(\neg a; b^{\smile}).$$

In the absence of boolean complement and converse, one has to define left and right semi inverses directly as operations on relations:

---

[1] *Journal of Logic, Language and Information*, 1992, vol. 3.

$R_a \backslash R_b \overset{\text{def}}{=} \{\langle x, y \rangle \mid \text{ for all } z \text{ with } \langle z, x \rangle \in R_a \text{ it holds that } \langle z, y \rangle \in R_b\},$

and similarly for the other semi-inverse.

These definitions give: $R_b \subseteq R_a; (R_a \backslash R_b)$ and $R_a \subseteq (R_a / R_b); R_b$.

In full relation algebra we can first leave out converse, and then see what we get by successively leaving out the operations of complement, star and residuation, in all possible orders.

Leave out star from full relation algebra and one gets the first order part of relation algebra (relation algebra proper, the fragment of first order logic with only two place relations). Leave out converse from this and one gets what Pratt calls residuated Boolean monoids (or residuated Boolean semirings).

Leave out converse from starred relation algebras and one gets Boolean action algebras. Leave out star from Boolean action algebras and one gets residuated Boolean monoids again. Leave out complement from Boolean action algebras and one gets action algebras. Leave out residuation from Boolean action algebras and one gets what Pratt calls Boolean starred semirings.

And so on, until we reach a system with only composition and union. Composition and union are the basic operations of the two dimensions. Structures closed under these operations are what Pratt calls idempotent semirings.

In fact, we can also move through the map via slightly different routes to discover new vistas. For instance, start with idempotent semirings and add residuation operators. This gives you residuated semirings. Now descend by leaving out union, and you will arrive at categorial logic or Lambek Calculus (which can be viewed as the logic of sequential composition).

Thus, we can get at action algebras, the logic which Pratt proposed a few years ago and now puts in its proper place with this roadmap, from below, from Lambek Calculus, as follows. First add union (the + operation), and then star.

The nice thing about the interplay of residuation and star is that the following principle holds: $a \backslash a = (a \backslash a)^*$. To see why this is so in the relational model, observe that $R_a \backslash R_a$ is defined as:

$\{\langle x, y \rangle \mid \text{ for all } z, \text{ if } \langle z, x \rangle \in R_a \text{ then } \langle z, y \rangle \in R_a\}.$

It follows immediately that $I \subseteq R_a \backslash R_a$ and that $R_a \backslash R_a; R_a \backslash R_a \subseteq R_a \backslash R_a$, so the relation is reflexive and transitive. It certainly is the smallest relation containing $R_a \backslash R_a$, so we get: $R_a \backslash R_a = (R_a \backslash R_a)^*$.

It seems to make excellent sense to add star to the Lambek Calculus. As far was we know, the resulting system was never thoroughly investigated. Still, it would seem to make an interesting topic of study, for this is in a sense the richest logic of concatenation.

The following table indicates that Pratt's roadmap covers areas which are to some extent *terra incognita*.

logic	decidability
Starred relation algebra	not decidable
Relation algebra	not decidable
Boolean action algebra	not decidable
Action algebra	open problem
KA	open problem
BSRT	open problem
RBM	undecidable
ISRT	open problem
RES	decidable
BSR	open problem
ISR	decidable
Lambek calculus	decidable

Note: This table was compiled with the help of Hajnal Andreka, Istvan Nemeti and Vaughan Pratt in December 1992. In July 1993, Andreka and Nemeti informed us that some of the terrain has been charted by further research in Budapest, and contributed a short chapter (written together with Ildiko Sain) with an overview of their results to this book.

Andreka, Nemeti and Sain   This paper gives an overview of decidability results for logics from Pratt's roadmap. It also poses the question of semantic completeness for these logics, and gives references to recent work with further results.

De Rijke   De Rijke presents a variant of dynamic logic which, like propositional dynamic lgic, makes a clear distinction between action and control. The difference with propositional dynamic logic is that the relations underlying the semantics of the modal operators are made visible at the syntactic level. To a certain extent this is already the case in propositional modal logic, of course, for the modality $\langle x \cup y \rangle$ shows that this is to be interpreted as the union of the relations for the modalities $\langle x \rangle$ and $\langle y \rangle$. But in dynamic modal logic modalities are more explicitly present, and the two sorts of propositions and relations are more closely linked.

In propositional dynamic logic, the only link from a program to a proposition is to a proposition expressing a postcondition of the program, and the only link in the other direction is from a proposition to a test for that proposition. In dynamic modal logic one can also go from a relation to its domain, its range, or its fixpoint, and from a proposition to the relations of expanding and contracting with that proposition (with respect to a fixed information ordering $\sqsubseteq$).

The connection with a suitably expanded version of propositional dynamic logic (with an operator for reversal, and an special relation $\sqsubseteq$ for making a move along the ordering of increase of information) is given by the following correspondences.

$do(\alpha) \rightsquigarrow \langle\alpha\rangle\top.$
$ra(\alpha) \rightsquigarrow \langle\alpha^{\smile}\rangle\top.$
$fix(\alpha) \rightsquigarrow \langle\alpha \cap \top?\rangle\top.$
$exp(\varphi) \rightsquigarrow \sqsubseteq;\varphi?.$
$contr(\varphi) \rightsquigarrow \sqsupseteq; \neg\varphi?.$

De Rijke shows that quite a number of dynamic or information logic can be embedded within dynamic modal logic. It turns out, however, that this versatility is bought at a price. Elsewhere, De Rijke has established the undecidability and in fact the high computational complexity of this logic. Arrow logic, presented in Van Benthem's chapter in this volume, was invented to get rid of some of the set theoretic assumptions causing this complexity.

Visser Visser's enterprise is akin to that of Gärdenfors, but whereas the Gärdenfors updates, at least those considered in his chapter in the present volume, do not change context, Visser considers updates that have a context change potential. More specifically, Visser considers the question of how to impose the additional structure on actions that is necessary for the handling of presuppositions.

There is a connection with Pratt and Kozen: Visser uses action logic as a tool. The general concern is as in the chapters of Van Eijck and Kracht: forging tools for natural language analysis.

The main line of Visser's chapter is as follows. We start with a suitable algebra representing the merging of informational items. (The specific choice of the chapter is to take a residuation lattice.) A condition can be isolated under which such an algebra can be viewed as representing update functions on a suitable set of states, the so called OTAT-principle. Here the states are considered as a special kind of informational items. Now we enrich the algebra by adding elements representing partial update functions. Such update functions are defined just in case the 'receiver' has a required amount of information available. It is shown that in this algebra the pieces of information required for convergence of the update functions behave as one would expect of decent presuppositions.

1.4 An Emerging Perspective?

Clearly, the time is not ripe for a unified theory of information, information linking, action and information flow. This books gives themes and variations, and some modulations connecting themes. Still, it is our hope that the volume will help to define the outlines of a perspective on information and information flow in which all these themes find their natural places in one surging symphony.

2 A Note on Dynamic Arrow Logic

Johan van Benthem

2.1 Finding the Computational Core

The current interest in logic and information flow has found its technical expression in various systems of what may be called 'dynamic logic' in some broad sense. But unfortunately, existing dynamic logics based on binary transition relations between computational states have high complexity (cf. Harel [6]). Therefore, it is worthwhile rethinking the choice of a relatively simple dynamic base system forming the 'computational core' that we need, without getting entangled in the complexity engendered by the additional 'mathematics of ordered pairs'. To some extent, this program is realized by various algebraic theories of programs and actions. But the conventional wisdom enshrined in these approaches may be biased: for instance, in its insistence that Boolean negation or complement is the main source of complexity. This may be seen by developing an alternative, namely a modal logic of 'arrows', which takes transitions seriously as dynamic objects in their own right (cf. van Benthem [3], Venema [19]). The main technical contribution of this chapter is the presentation of a system of Arrow Logic with both first-order relational operations and infinitary Kleene iteration, which may be a good candidate for the computational core calculus. In particular, we prove completeness for its minimal version, and establish various connections with propositional dynamic logic and categorial logics.

There is a more general program behind the above proposal. For instance, one can do the same kind of 'arrow analysis' for many other systems in the computational literature, such as the 'dynamic modal logic' in De Rijke's contribution to this volume. Moreover, issues of apparent undecidability and higher-order complexity abound in the semantics of programming. For instance, in Hoare Logic, infinitary control structures create high complexity: is this inevitable, or can the situation be mitigated by redesign? Likewise, in Knowledge Representation, higher-order data structures (such as 'branches' in trees) can generate complexity, witness the field of 'branching temporal logic', which may be avoided by suitable re-analysis in many-sorted first-order theories. Thus, the general issue raised in this chapter is the following:

What is genuine 'computation' and what is 'extraneous mathematics' in the logical analysis of programming and its semantics?

If we can isolate the former component, many different technical results in the current literature might be separable into different computational content plus a repetition of essentially the same mathematical overhead. We do not offer any general solution to this question here, but we do advocate some general awareness of the phenomenon.

2.2 Arrow Logic in a Nutshell

The intuition behind Arrow Logic is as follows. Binary relations may be thought of as denoting sets of arrows. Some key examples are 'arcs' in graphs, or 'transitions' for dynamic procedures in Computer Science, but one can also think of 'preferences' in the case of ranking relations (as found in current theories of reasoning in Artificial Intelligence, or social choice and economics). These arrows may have internal structure, whence they need not be identified with ordered pairs (source, target): several arrows may share the same input-output pair, but also certain pairs may not be instantiated by an arrow. This motivates the following definitions (what follows here is a brief sketch: we refer to the references in the text for further technical details):

Arrow Frames are tuples (A, C^3, R^2, I^1) with
$\quad A \qquad$ a set of objects ('arrows') carrying three predicates:
$\quad C^3 x, yz \quad$ x is a 'composition' of y and z
$\quad R^2 x, y \quad$ y is a 'reversal' of x
$\quad I^1 x \qquad$ x is an 'identity' arrow

Arrow Models M add a propositional valuation V here, and one can then interpret an appropriate modal propositional language expressing properties of (sets of) arrows using two modalities reflecting the basic 'ordering operations' of relational algebra:

$$
\begin{aligned}
&M, x \models p && \text{iff} && x \in V(p) \\
&M, x \models \neg\varphi && \text{iff} && \textit{not } M, x \models \varphi \\
&M, x \models \varphi \wedge \psi && \text{iff} && M, x \models \varphi \textit{ and } M, x \models \psi \\
&M, x \models \varphi \bullet \psi && \text{iff} && \textit{there exist } y, z \textit{ with } C\,x, yz \textit{ and} \\
& && && M, y \models \varphi,\ M, z \models \psi \\
&M, x \models \varphi^{\vee} && \text{iff} && \textit{there exists } y \textit{ with } R\,x, y \textit{ and } M, y \models \varphi \\
&M, x \models Id && \text{iff} && I\,x
\end{aligned}
$$

The minimal modal logic of this system is an obvious counterpart of its mono-modal predecessor, whose key principles are axioms of Modal Distribution:

$$
\begin{aligned}
(\varphi_1 \vee \varphi_2) \bullet \psi &\leftrightarrow (\varphi_1 \bullet \psi) \vee (\varphi_2 \bullet \psi) \\
\varphi \bullet (\psi_1 \vee \psi_2) &\leftrightarrow (\varphi \bullet \psi_1) \vee (\varphi \bullet \psi_2) \\
(\varphi_1 \vee \varphi_2)^{\vee} &\leftrightarrow \varphi_1^{\vee} \vee \varphi_2^{\vee}
\end{aligned}
$$

A completeness theorem is provable here along standard lines, using Henkin models.

(This minimal logic includes all the usual laws of Boolean Algebra.)

Next, one can add further axiomatic principles (taking cues from relational algebra) and analyze what constraint these impose on arrow frames via the usual semantic correspondences. In particular, we have that

$$\neg(\varphi)^\vee \to (\neg\varphi)^\vee \quad \text{iff} \quad \forall x\, \exists y\, R\,x,y \tag{2.1}$$

$$(\neg\varphi)^\vee \to \neg(\varphi)^\vee \quad \text{iff} \quad \forall xyz : (R\,x,y \wedge R\,x,z) \to y = z \tag{2.2}$$

Together, these make the binary relation R into a unary function r of 'reversal'. Then the 'double conversion' axiom makes the function r idempotent:

$$(\varphi)^{\vee\vee} \leftrightarrow \varphi \quad \text{iff} \quad \forall x\, r(r(x)) = x. \tag{2.3}$$

Let us assume this much henceforth in our arrow frames. Next, the following principles of Relational Algebra regulate the interaction of reversal and composition:

$$(\varphi \bullet \psi)^\vee \to \psi^\vee \bullet \varphi^\vee \quad \text{iff} \quad \forall xyz : C\,x,yz \to C\,r(x), r(z)r(y) \tag{2.4}$$

$$\varphi \bullet \neg(\varphi^\vee \bullet \psi) \to \neg\psi \quad \text{iff} \quad \forall xyz : C\,x,yz \to C\,z, r(y)x \tag{2.5}$$

Together (2), (4), (5) imply the further interchange law

$$\forall xyz : C\,x,yz \to C\,y, xr(z).$$

Moreover, there is actually a more elegant form of axiom (5) without negation:

$$\varphi \wedge (\psi \bullet \chi) \to \psi \bullet (\chi \wedge (\psi^\vee \bullet \varphi)).$$

Finally, the propositional constant Id will be involved in correspondences like

$$Id \to Id^\vee \quad \text{iff} \quad \forall x : I\,x \to I\,r(x) \tag{2.6}$$

$$Id \bullet \varphi \to \varphi \quad \text{iff} \quad \forall xyz : (I\,y \wedge C\,x,yz) \to x = z. \tag{2.7}$$

Obviously, there are many further choices here, and 'Arrow Logic' really stands for a family of modal logics, whose selection may depend on intended applications. Nevertheless, what might be the most natural 'computational core' in this field? Our recommendation would be as follows:

Universal frame Constraints

Take only those principles concerning composition, converse and identity on arrow frames which lack existential import: i.e., their corresponding constraints can be formulated by purely *universal* first-order sentences.

One potential exception to this proposal is Associativity for composition:

$$\text{a. } (\varphi \bullet \psi) \bullet \chi \to \varphi \bullet (\psi \bullet \chi) \quad \text{iff} \quad \forall xyzuv : (C\,x,yz \wedge C\,y,uv) \to \tag{2.8}$$
$$\exists w : (C\,x,uw \wedge C\,w,vz))$$

b. and likewise in the opposite direction.

Associativity is often tacitly presupposed in the formulation of dynamic semantics. In what follows we shall avoid this practice.

Further information about the landscape of systems for Arrow Logic may be found in van Benthem [3], Marx, Németi and Sain [7], Vakarelov [18], and Venema [19] (cf. also Appendix 2A). Two further technical points deserve mention here. One is the existence of a certain uniformity. The above correspondences all follow from a general result in Modal Logic called the *Sahlqvist Theorem* in van Benthem [2], which supplies an algorithm for computing frame conditions that can also be applied to other proposed candidates for inclusion in our core set. The other point is that the present modal language also has clear limits to its expressive power. Notably, one cannot force composition to become a partial function (general arrow logic allows more than one way of composing two transitions). For the latter purpose, *enriched modal formalisms* will be needed, employing further modal operators (cf. de Rijke [14], Roorda [16]), which we shall not pursue here. Of course, by the time we have enforced full representability of arrow frames via sets of ordered pairs, the resulting modal logic will be just as complex as the ordinary theory of representable relational algebras. The art is to know when to stop.

2.3   A Complete System of Dynamic Arrow Logic

Now, *Dynamic Arrow Logic* adds one infinitary operator to the above language:

$M, x \models \varphi^*$ iff x *can be C-decomposed into some finite sequence of arrows satisfying φ in M.*

What this says is that there exists some finite sequence of φ-arrows in M which allows of at least one way of successive composition via intermediate arrows so as to arrive at x. (Without Associativity, this does not imply that x could be obtained by any other route of combinations from these same arrows.) Intuitively, φ^* describes the transitive closure of φ. It satisfies the following simple and natural principles:

$$axiom \quad \varphi \to \varphi^* \tag{2.9}$$

$$axiom \quad \varphi^* \bullet \varphi^* \to \varphi^* \tag{2.10}$$

$$rule \quad if \varphi \to \alpha \ and \ \alpha \bullet \alpha \to \alpha \ are \ provable, \ then \tag{2.11}$$
$$so \ is \ \varphi^* \to \alpha$$

These principles may be added to the earlier minimal arrow logic, to obtain a simple base system, but our preferred choice will consist of this minimal basis plus the earlier principles (1)–(5), to obtain a suitable axiomatic Dynamic Arrow Logic DAL. Here is an illustration of how this system works.

EXAMPLE 2.1   Derivation of Monotonicity for Iteration

If $\vdash \alpha \to \beta$ $\qquad\qquad$ then $\vdash \alpha \to \beta^*$ (axiom (9))
Also $\vdash \beta^* \bullet \beta^* \to \beta^*$ (axiom (10))$\quad$ whence $\vdash \alpha^* \to \beta^*$ (rule (11))$\qquad$ ∎

EXAMPLE 2.2   Derivation of Interchange for Iteration and Converse

i	$\varphi \to \varphi^*$	axiom (9)
ii	$\varphi^\vee \to \varphi^{*\vee}$	i plus monotonicity for converse (the latter follows from its distributivity)
iii	$(\varphi^{*\vee} \bullet \varphi^{*\vee}) \to (\varphi^* \bullet \varphi^*)^\vee$	this may be derived using axioms (3) and (4)
iv	$\varphi^* \bullet \varphi^* \to \varphi^*$	axiom (10)
v	$(\varphi^* \bullet \varphi^*)^\vee \to \varphi^{*\vee}$	iv plus monotonicity for converse
vi	$(\varphi^{*\vee} \bullet \varphi^{*\vee}) \to \varphi^{*\vee}$	iii, v
vii	$\varphi^{\vee *} \to \varphi^{*\vee}$	ii, vi plus rule (11)
viii	$\varphi^{\vee\vee *} \to \varphi^{\vee * \vee}$	by similar reasoning
ix	$\varphi \to \varphi^{\vee\vee}$	axiom (3)
x	$\varphi^* \to \varphi^{\vee\vee *}$	monotonicity for iteration
xi	$\varphi^* \to \varphi^{\vee * \vee}$	x, viii
xii	$\varphi^{*\vee} \to \varphi^{\vee * \vee\vee}$	xi plus monotonicity for converse
xiii	$\varphi^{\vee * \vee\vee} \to \varphi^{\vee *}$	axiom (2)
xiv	$\varphi^{*\vee} \to \varphi^{\vee *}$	xii, xiii $\qquad$ ∎

Completeness may be established for DAL, as well as several of its variants.

THEOREM 2.1   DAL is complete for its intended interpretation.

Proof   Take some finite universe of relevant formulas which is closed under subformulas and which satisfies the following closure condition:

if φ^* is included, then so is $\varphi^* \bullet \varphi^*$

Now consider the usual model of all maximally consistent sets in this restricteduniverse, setting (for all 'relevant' formulas):

$C\,x,yz$ $\quad$ iff $\quad \forall \varphi \in y, \psi \in z : \varphi \bullet \psi \in x$
$R\,x,y$ $\quad$ iff $\quad \forall \varphi \in y : \varphi^\vee \in x$

Here we can prove the usual 'decompositions' for maximally consistent sets, such as

$\varphi \bullet \psi \in x$ iff there exist y, z with $C\,x,yz$ and $\varphi \in y, \psi \in z$

using the minimal distribution axioms only. The key new case here is the following:

CLAIM 2.1 $\varphi^* \in x$ iff some finite sequence of maximally consistent sets each containing φ 'C-composes' to x in the earlier sense.

Proof From right to left. This is a straightforward induction on the length of the decomposition, using axioms (9), (10) and the closure condition on relevant formulas.

From left to right. Describe the finite set of all 'finitely C-decomposable' maximally consistent sets in the usual way by means of one formula α, being the disjunction of all their conjoined 'complete descriptions' δ. Then we have

$$\vdash \varphi \rightarrow \alpha :$$

since φ is provably equivalent to $\bigvee_{\varphi \in \delta} \delta$ in propositional logic, and α contains all these δ by definition. Next, we have

$$\vdash \alpha \bullet \alpha \rightarrow \alpha :$$

To see this, suppose that $(\alpha \bullet \alpha) \wedge \neg\alpha$ were consistent. Using Distributivity with respect to successive relevant formulas, $(\delta_1 \bullet \delta_2) \wedge \neg\alpha$ must be consistent for some maximally consistent δ_1, δ_2. And likewise, $(\delta_1 \bullet \delta_2) \wedge \neg\alpha \wedge \delta_3$ will be consistent for some maximally consistent δ_3. Now, δ_1, δ_2 must be in α, and moreover $C\, \delta_3, \delta_1\delta_2$ by the definition of C and some deductive reasoning. Therefore, δ_3 is in α too (by definition), contradicting the consistency of $\neg\alpha \wedge \delta_3$. So, applying the iteration rule (11), we have

$$\vdash \varphi^* \rightarrow \alpha$$

Therefore, if $\varphi^* \in x$, then x belongs to α.

Semantic evaluation in the canonical model will now proceed in harmony with the above syntactic decomposition: any relevant formula is true 'at' a maximally consistent set iff it belongs to that set. This completes our analysis of the basic case.

In order to deal with the additional axioms (1)–(5), their frame properties must be enforced in our finite canonical model. This may be done as follows:

i.   one closes the universe of relevant formulas under Boolean operations and converses: the resulting infinite set of formulas will remain logically finite;

ii.   given the Boolean laws and the interchange principles for converse, the definition of the relation C is to be modified by adding suitable clauses, so as to 'build in' the required additional frame properties.

First, the required behaviour of reversal is easy to obtain. One may define $r(x)$ to be the maximally consistent set consisting of (all representatives of) $\{\varphi^\vee \mid \varphi \in x\}$: the available axioms make this an idempotent function inside the universe of relevant maximally consistent sets. For a more difficult case, consider axiom (5) with corresponding frame condition $\forall xyz : C\, x, yz \rightarrow C\, z, r(y)x$. One redefines:

$C\,x,yz$ iff both $\forall\varphi\in y, \psi\in z : \varphi\bullet\psi\in x$

and $\forall\varphi^\vee\in y, \psi\in x : \varphi\bullet\psi\in z$

This has been designed so as to validate the given frame condition. But now, we need to check that the earlier decomposition facts concerning maximally consistent sets are still available, to retain the harmony between membership and truth at such sets. Here are the two key cases:

$\varphi\bullet\psi\in x$ iff there exist y, z with $C\,x,yz$ and $\varphi\in y$, $\psi\in z$

$\varphi^*\in x$ iff some finite sequence of maximally consistent sets containing φ 'C–composes' to x.

The crucial direction here is from left to right: can we find maximally consistent sets as required with C satisfying the additional condition? What we need is this. In the earlier proof, the sets y, z were constructed 'globally', by showing how successive selection yields a consistent set of formulas x, $\bigwedge y\bullet\bigwedge z$ with $\varphi\in y$, $\psi\in z$. (For then, whenever $\alpha\bullet\beta$ is a relevant formula with $\alpha\in y$, $\beta\in z$, $\alpha\bullet\beta$ must belong to x, on pain of inconsistency.) Now, it suffices to show that, in this same situation, the set z, $\bigwedge r(y)\bullet\bigwedge x$ is consistent too. Here, we use a rule derived from axioms (2), (5):

if $\vdash \varphi\to\neg(\psi\bullet\chi)$ then $\vdash \chi\to\neg(\psi^\vee\bullet\varphi)$:

Then, if $\vdash z\to\neg(\bigwedge r(y)\bullet\bigwedge x)$, then $\vdash x\to\neg(\bigwedge rr(y)\bullet\bigwedge z)$, and hence also $\vdash x\to$ $\neg(\bigwedge y\bullet\bigwedge z)$: contradicting the consistency of x, $\bigwedge y\bullet\bigwedge z$. The argument for iteration is similar. Moreover, the general case with all frame conditions implanted simultaneously employs the same reasoning. ∎

COROLLARY 2.1   DAL is decidable.

Proof The preceding argument establishes not just axiomatic completeness but also the finite Model Property. ∎

The above strategy for accommodating the relevant additional frame properties in the finite counter-model is that of Roorda [15]. More generally, we conjecture that every modal logic which is complete with respect to some Finite set of Horn clause frame conditions has the Finite Model Property. But, will decidability go through if the further existential property of Associativity is included in our basic arrow logic? This is more difficult, since the required additional worlds, whose existence is easily shown in a full Henkin model by traditional arguments, seem to over-flow the finite universe of 'relevant' maximally consistent sets during filtration. Dimiter Vakarelov has announced a proof, but the negative results in Andréka [1] counsel caution.

Another strategy uses a *labeled version* of arrow logic with statements of the form 'arrow : assertion', which transcribe the above truth definition into a simple fragment of predicate logic. Labels might be either bare arrows (with relations of composition and converse), or complex descriptions from some (semi-)group. Following Roorda [16], decidability might then be proved via an effective equivalence with some cut-free labeled sequent calculus for Arrow Logic, whose rules might use a format like:

$$\Sigma, x : A, y : B \vdash \Delta \quad \text{implies} \quad \Sigma, xy : A \bullet B \vdash \Delta$$
$$\Sigma \vdash \Delta, r(x) : A \quad \text{implies} \quad \Sigma \vdash \Delta, x : A^\vee$$

2.4   Propositional Dynamic Logic with Arrows

Now what about a *Propositional Dynamic Logic* based on the above? The usual account in the literature considers the addition of a propositional component referring to truth at states essential, as it allows us some negation at least at the latter level. Since this is no longer true now, having this second component becomes more of a convenience. Nevertheless, we do think the resulting two-level system is a natural one: 'arrow talk' and 'state talk' belong together in an analysis of computation and general action. So as usual, add a Boolean propositional language, plus two mechanisms of interaction between the two resulting components:

> a *test* 'mode' ? taking statements to programs
> a *domain* 'projection' $\langle\rangle$ taking programs to statements.

For notational convenience, we shall reserve $\varphi, \psi, \ldots$ henceforth for state assertions and $\pi, \pi_1, \pi_2, \ldots$ for describing programs in this two-tier system.

In line with the general modal analysis of the above, let us view this system with some greater abstraction. What we have is a *two-sorted modal logic*, whose models have both 'states' and 'arrows', and whose formulas are marked for intended interpretation at one of these. Both the arrow and state domains may carry internal structure, reflected in certain modalities, such as the earlier $\bullet$ and $^\vee$ referring to arrows. (States might be ordered by 'precedence' or 'preference' with appropriate modalities.) Our key point, however, is this. Even the modes and projections themselves may be viewed as 'non-homogeneous' modalities, reflecting certain structure correlating the two kinds of object in our models. For instance, 'test' is again a *distributive* modality, and so is 'domain':

$$(\varphi \vee \psi)? \quad \leftrightarrow \quad \varphi? \vee \psi?$$
$$\langle \pi_1 \vee \pi_2 \rangle \quad \leftrightarrow \quad \langle \pi_1 \rangle \vee \langle \pi_2 \rangle$$

whose interpretations run as follows:

$$M, x \models \varphi? \quad \text{iff} \quad \textit{there exists some } s \text{ with } Tx, s \text{ and } M, s \models \varphi$$
$$M, x \models \langle \pi \rangle \quad \text{iff} \quad \textit{there exists some } x \text{ with } Ds, x \text{ and } M, x \models \pi$$

Intuitively, the first relation Tx, s says that x is an identity arrow for the point s, while the second relation Ds, x says that s is a left end-point of the arrow x.

Via the usual correspondences, further axioms on ?, $\langle \rangle$ will then impose additional connections between T and D.

EXAMPLE 2.3   Connecting Identity Arrows and End-Points

The principle $\langle \varphi ? \rangle \leftrightarrow \varphi$ (itself again a modal 'Sahlqvist form') expresses the conjunction of

$$\forall s \exists x : Ds, x \wedge Tx, s \quad \forall sx : Ds, x \rightarrow \forall s' : Tx, s' \rightarrow s = s'. \qquad \blacksquare$$

Also, axiomatic completeness proofs are straightforward here, with two kinds of maximally consistent sets: one for arrows and one for points. Thus everything about Propositional Dynamic Logic is Modal Logic: not just its two separate components, but also their connections.

Further elegance may be achieved here by a reformulation. The following observation is made in van Benthem [3]:

FACT 2.1   There is one projection which is a Boolean homomorphism, namely the diagonal function $\lambda R \bullet \lambda x \bullet Rxx$. There are exactly two homomorphic modes, namely $\lambda P \bullet \lambda xy \bullet Px$ and $\lambda P \bullet \lambda xy \bullet Py$.

Thus, we can introduce three matching modalities with corresponding new binary relations in their semantics:

$$
\begin{array}{lll}
M, s \models D\pi & \text{iff} & \text{for some } x, \Delta s, x \text{ and } M, x \models \pi \\
M, x \models L\varphi & \text{iff} & \text{for some } s, \mathcal{L}s, x \text{ and } M, s \models \varphi \\
M, x \models R\varphi & \text{iff} & \text{for some } s, \mathcal{R}s, x \text{ and } M, s \models \varphi
\end{array}
$$

These modalities satisfy not just the Distribution axioms, but they also commute with Boolean negation (just like relational converse), so that we can take $\Delta, \mathcal{L}, \mathcal{R}$ to be *functions*. This set-up is more elegant, as well as easy to use. (It may still be simplified a bit by dropping $R\varphi$ in favour of $(L\varphi)^{\vee}$.) For instance, one source of axiomatic principles is the interaction of various operators:

OBSERVATION 2.1

$$
\begin{array}{lll}
DL\varphi \leftrightarrow \varphi & \text{expresses that} & \forall s : \mathcal{L}\Delta(s) = s \\
LD\pi \leftrightarrow (\pi \wedge Id) \bullet T & \text{expresses that} & \forall x \exists y : Cx, \Delta \mathcal{L}(x)y \wedge \\
& & \forall xyz : \left(\begin{array}{c} (Cx, yz \wedge Iy) \\ \rightarrow y = \Delta \mathcal{L}(x) \end{array} \right) \\
L\varphi \bullet \pi \rightarrow L\varphi & \text{expresses that} & \forall xyz : Cx, yz \rightarrow \mathcal{L}(x) = \mathcal{L}(y).
\end{array}
$$

One may achieve exactly the power of the standard system with these new primitives under the following

Translation from old to new format

$$\varphi? : L\varphi \wedge Id \qquad \langle\pi\rangle : D(\pi \bullet T)$$

Analyzing the usual axioms of Propositional Dynamic Logic in this fashion is a straightforward exercise. We list the key principles that turn out to be needed (these allow us to represent statements $\langle\pi\rangle\varphi$ faithfully as $D((\pi \wedge R\varphi) \bullet T)$:

1   $D\pi \to D(\pi \wedge Id)$

2   $Id \to (L\varphi \leftrightarrow R\varphi)$

3a $DL\varphi \leftrightarrow \varphi$

3b $DR\varphi \leftrightarrow \varphi$

4   $\pi_1 \wedge RD\pi_2 \leftrightarrow \pi_1 \bullet (\pi_2 \wedge Id)$

5   $(\pi_1 \bullet \pi_2) \wedge R\varphi \leftrightarrow \pi_1 \bullet (\pi_2 \wedge R\varphi)$

Their corresponding frame conditions can be computed by hand, or again with a Sahlqvist algorithm, as they are all of the appropriate modal form. These principles suffice for deriving various other useful ones, such as the reductions

$$(Id \wedge L\varphi) \bullet \pi \leftrightarrow L\varphi \wedge \pi \qquad \pi \bullet (Id \wedge R\varphi) \leftrightarrow \pi \wedge R\varphi.$$

Finally, there is also a converse route, via two more schemata:

Translation from old to new format

$$L\varphi : \varphi? \bullet T \qquad D\pi : \langle Id \wedge \pi\rangle$$

The same style of analysis may be applied to richer systems of dynamic logic, having additional structure in their state domains (cf. van Benthem [3], de Rijke [13]). One example is the 'dynamic modal logic' in De Rijke's contribution to this Volume, which features modes over information states with an inclusion order $\subseteq$. This may be treated by introducing another propositional constant at the arrow level, say, E for 'inclusion' (perhaps with suitable axioms expressing its transitivity and reflexivity). Then, the logic of updating and revision will employ special defined arrows, such as

$E \wedge R\varphi$ update transition for φ

$(E \wedge R\varphi) \wedge \neg((E \wedge R\varphi) \bullet (E \wedge \neg Id))$ minimal update transition

                                                    for φ

This may provide a workable alternative where the undecidability of the full system is circumvented. Roughly speaking, the arrow version should stay on the right side of the '2D-boundary' which allows embedding of two-dimensional grids in the models, and hence encoding of full Turing machine computation.

Acknowledgements

I would like to thank Hajnal Andréka, Istvan Németi, Maarten de Rijke, Ildikó Sain, and Dimiter Vakarelov for their helpful comments on this draft, as well as their general response to 'Arrow Logic'.

Appendixes

2A   From Amsterdam to Budapest

Arrow Logic in its 'Amsterdam manifestation' says that dynamic transitions need not be identified with the ordered pairs over some underlying state set. This idea has really two different aspects. Distinct arrows may correspond to the same pair of ⟨input, output⟩, but also, not every such pair need correspond to an available arrow. This shows very well in the following less standard example:

Let arrows be *functions* $f : A \to B$ giving rise to, but not identifiable with, ordered pairs $\langle A, B \rangle$ of 'source' and 'target'. Then, the relation C expresses the partial function of composition of mappings, while the reversal relation R will hold between a function and its inverse, if available.

This model will validate all of the earlier core principles, at least, in their appropriate versions after functionality for reversal has been dropped. For instance, axiom (5) now expresses the fact that, whenever $f = g \circ h$ and $k = g^{-1}$, then also $h = k \circ f$.

Nevertheless, there is also an interesting more 'conservative' variant found in various earlier and recent publications from Budapest, where arrows are still ordered pairs, but one merely gives up the idea that all ordered pairs are available as arrows. Essentially this takes us to a universally first-order definable class of arrow frames which can be represented via sets of ordered pairs (though not necessarily full Cartesian products). Its complete logic can be determined in our formalism, and it turns out to be decidable as well (Marx, Németi and Sain [7]). This system is another natural, richer stopping point in the arrow landscape, including the earlier systems presented in Section 2.2 above, with additional axioms expressing essentially the uniqueness of the pair of identity arrows surrounding an arbitrary arrow, as well as their 'proper fit' with composition and reversal.

Various weaker natural arrow logics with desirable meta-properties (decidability, interpolation, etcetera) may be found in Németi [9] (see also the survey Németi [10] for more extensive documentation). Simón [17] investigates deduction theorems for arrow logics, showing that our basic systems lacks one. Finally, Andréka [1] provides a method for proving results on non-finite-axiomatizability in the presence of full Associativity. Even undecidability lies around the corner, in this perspective, as soon as one acquires enough power to perform the usual encoding of relation-algebraic quasi-equations into equations.

2B   Dynamic Arrow Logic with a Fixed-Point Operator

The analysis of this chapter may be extended to prove completeness for a more powerful system of Dynamic Arrow Logic which has the well-known minimal fixed-point operator $\mu p \cdot \varphi(p)$. Its two key derivation rules are as follows:

$$if \vdash \varphi(\alpha) \to \alpha \qquad then \vdash \mu p \cdot \varphi(p) \to \alpha \qquad (I)$$
$$if \vdash \beta \to \mu p \cdot \varphi(p) \quad then \vdash \varphi(\beta) \to \mu p \cdot \varphi(p) \qquad (II)$$

This language defines iterations φ^* in our sense via the fixed-point formula
$\mu p \cdot \varphi \vee p \bullet p.$

(Its successive approximations give us all C-combinations that were involved in the earlier semantic definition.) The derivation rules for iteration then become derivable from the above two rules: I corresponds to rule (11), while II gives the effect of the axioms (9), (10). In the completeness theorem, these allow us to generalize the earlier argument for the crucial decomposition:

$\mu p \cdot \varphi(p) \in x$ iff x belongs to some finite iteration of the operator $\lambda p \cdot \varphi(p)$ starting from the empty set for p.

2C   Connections with Categorial Logic and Action Algebra

Dynamic Arrow Logic may also be compared to a dynamic version of categorial logic, as employed in current categorial grammars, extended with Kleene iteration. At the base level, this connection runs between ordinary arrow logic and standard systems such as the Lambek Calculus with two directed functional slashes (cf. van Benthem [3, 4] for details):

$$a \backslash b := \neg(a^\vee \bullet \neg b) \qquad b/a := \neg(\neg b \bullet a^\vee)$$

Moreover, categorial product goes to composition $\bullet$. The two basic categorial laws then express the basic interaction principles for C and r on arrow frames:

$a \bullet (a \backslash b) \le b \qquad \forall xyz : Cx, yz \to Cz, r(y)x$

$(b/a) \bullet a \le b \qquad \forall xyz : Cx, yz \to Cy, xr(z)$.

This gives us the two implications

$X \le a \backslash b \Rightarrow a \bullet X \le b \qquad X \le b/a \Rightarrow X \bullet a \le b$

Their converses (which generate all of the Lambek Calculus) require no more. For instance, suppose that $a \bullet X \le b$. Now, $X \wedge (a^\vee \bullet \neg b) \le a^\vee \bullet (\neg b \wedge (a \bullet X))$ (by the first interaction principle), whence $X \wedge (a^\vee \bullet \neg b) \le a^\vee \bullet (\neg b \wedge b)$ and then $X \wedge (a^\vee \bullet \neg b) \le a^\vee \bullet 0 \le 0$. I.e., $X \le \neg(a^\vee \bullet \neg b)$.

Thus, Basic Arrow Logic contains the Lambek Calculus, and it even does so faithfully, thanks to the completeness theorem in Mikulás [8]. Many further connections between categorial logics and arrow logics remain to be explored.

With $*$ added, we get some obvious further principles, such as $(a \backslash a)^* = (a \backslash a)$, due to Tarski and Ng. Note how this may be derived in Dynamic Arrow Logic:

1 $(a \backslash a) \le (a \backslash a)^*$ axiom 9

2 $(a \backslash a) \le (a \backslash a)$

  $(a \backslash a) \bullet (a \backslash a) \le (a \backslash a)$ by the above categorial rules

                                    (derivable in arrow logic)

  $(a \backslash a)^* \le (a \backslash a)$ by rule (11)

A related system is the Action Algebra of Pratt [11] and previous publications, which may be viewed as a standard categorial logic enriched with iteration and disjunction. It would be of interest to determine the precise connection with arrow logic here. What is easy to determine, at least, is the following 'arrow content' of the equational axiomatization offered by Pratt. Its basic axioms each exemplify one of four kinds of assertion in our framework:

1. consequences of the minimal arrow logic – in particular, the basic laws of monotonicity (a typical example is "$a \to b \le a \to (b + b')$")

2. expressions of categorial principles, whose content was the basic interaction between composition and converse (as expressed in the inequalities "$a(a \to b) \le b \le a \to ab$")

3. universally valid principles for iteration, such as its monotonicity (compare "$a^* \le (a + b)^*$")

4. associativity for composition (whose precise strength remains to be determined in the arrow framework, as we have seen).

2D   Predicate Arrow Logic

It may also be of interest to ask whether the above style of analysis applies to ordinary predicate logic. In particular, does its undecidability go away too, once we give up the usual bias toward ordered pairs? First, the formulation is easy:

Take a two-sorted language with 'objects' and 'arrows' and read, say, "Rxy" as $\exists a (Ra \wedge l(a) = x \wedge r(a) = y)$.

Thus we need unary predicates for the old relations, plus two new auxiliary cross- sorted maps l, r identifying end-points of arrows. (For general n-any relations, we may need a more-dimensional version of Arrow Logic, as in Vakarelov [18].) But the resulting system still faithfully embeds ordinary predicate logic, and hence it is at least as complex.

QUESTION 2.1   What would have to be weakened in standard predicate logic to get an arrow-based decidable version, either in the Amsterdam or the Budapest Way?

What this analysis shows is that versions of Arrow Logic can also get undecidable without identifying arrows with ordered pairs, viz. by putting in additional expressive power via modal operators reflecting further predicate-logical types of statement, such as a 'universal modality' or a 'difference operator' or yet other additions (cf. again de Rijke [14], Roorda [16] – as well as the various recent publications by Gargov, Goranko, Passy, and others from the 'Sofia School' in enriched modal logic).

QUESTION 2.2   What happens to the previous versions of Propositional Arrow Logic if one adds a 'universal modality' or a 'difference operator', or yet other notions from extended Modal Logic?

A good concrete example here is the traditional formula enforcing infinity of a binary relation in its models:

$$\forall x \neg Rxx \wedge \forall x \exists y\, Rxy \wedge \forall xy\, (Rxy \rightarrow \forall z(Ryz \rightarrow Rxz)).$$

Its 'arrow transcription' reflects our natural reasoning about this formula, in terms of growing chains in arrow diagrams. Analyzing the usual argument about its models, one finds how little is needed to show their infinity, thus destroying the finite Model Property and endangering Decidability.

References

[1] H. Andréka, 1991. 'Representations of Distributive Semilattice-Ordered Semigroups with Binary Relations', *Algebra Universalis* 28, 12-25.

[2] J. van Benthem, 1984. 'Correspondence Theory', in D. Gabbay and F. Guenthner, eds., 167-247.

[3] J. van Benthem, 1991. *Language in Action. Categories, Lambdas and Dynamic Logic*, Elsevier Science Publishers, Amsterdam, (Studies in Logic, vol. 130).

[4] J. van Benthem, 1992. 'Logic and the flow of Information', in D. Prawitz et al., eds., to appear.

[5] D. Gabbay and F. Guenthner, eds, 1984. *Handbook of Philosophical Logic*, vol. II, Reidel, Dordrecht.

[6] D. Harel, 1984. 'Dynamic Logic', in D. Gabbay and F. Guenthner, eds., 497-604.

[7] M. Marx, I. Németi and I. Sain, 1992. 'Everything You Always Wanted to Know about Arrow Logic', Center for Computer Science in Organization and Management, University of Amsterdam / Mathematical Institute of the Hungarian Academy of Sciences, Budapest.

[8] S. Mikulás, 1992. 'Completeness of the Lambek Calculus with respect to Relational Semantics', Research Report LP-92-03, Institute for Logic, Language and Computation, University of Amsterdam.

[9] I. Németi, 1987. 'Decidability of Relation Algebras with Weakened Axioms for Associativity', *Proceedings American Mathematical Society* 100:2, 340-345.

[10] I. Németi, 1991. 'Algebraizations of Quantifier Logics, An Introductory Overview', to appear in *Studia Logica*, special issue of Quantifier Logic (W. Blok and D. Pigozzi, eds.).

[11] V. Pratt, 1992. 'Action Logic and Pure Induction', *this Volume*.

[12] D. Prawitz, B. Skyrms and D. Westerståhl, eds, to appear. *Proceedings 9th International Congress for Logic*, Methodology andPhilosophy of Science. Uppsala 1991, North-Holland, Amsterdam.

[13] M. de Rijke, 1992. 'A System of Dynamic Modal Logic', Report LP-92-08, Institute for Logic, Language and Computation, University of Amsterdam.

[14] M. de Rijke, 1992. 'The Modal Logic of Inequality', *Journal of Symbolic Logic* 57:2, 566-584.

[15] D. Roorda, 1991. *Resource Logics. Proof-Theoretical Investigations*, Dissertation, Institute for Logic, Language and Computation, University of Amsterdam.

[16] D. Roorda, 1992. *Lambek Calculus and Boolean Connectives: On the Road*, Onderzoeksinstituut voor Taal en Spraak, Rijksuniversiteit, Utrecht.

[17] A. Simon, 1992. 'Arrow Logic Lacks the Deduction Theorem', Mathematical Institute of the Hungarian Academy of Sciences, Budapest.

[18] D. Vakarelov, 1992. 'A Modal Theory of Arrows I', Report ML-92-04, Institute for Logic, Language and Computation, University of Amsterdam.

[19] Y. Venema, 1992. *Many-Dimensional Modal Logic*, Dissertation, Institute for Logic, Language and Computation, University of Amsterdam.

3 Axiomatizing Dynamic Predicate Logic with Quantified Dynamic Logic

Jan van Eijck

3.1 Introduction

Dynamic predicate logic was proposed in [8] as a medium for natural language meaning representation that is able to deal with unselective binding of pronouns. This logic can be axiomatized with tools from Hoare logic [4]; in a Hoare style approach, static truth conditions for dynamic programs are calculated as weakest preconditions for success of those programs. This chapter investigates several notions of entailment for dynamic predicate logic and proposes a more elegant axiomatization of dynamic predicate logic. It is shown that the rules of the Hoare style axiomatization arise as derived inference rules in our more comprehensive axiomatisation which uses Quantified Dynamic Logic in the style of Pratt [11].

Dynamic predicate logic (DPL) is a tool for the representation of knowledge conveyed by means of natural language. Due to limitations of space we will not dwell on a detailed analysis of the use of DPL as an NL knowledge representation tool (see [8] for that). Suffice it to say that DPL extended with definite assignments can be used to represent the meanings of natural language sentences such as (3.1). (I wish to apologize in advance to all those who think the following example is not politically correct. People who know me can vouch for the fact that I am not a sexist.)

If a woman is married, then her husband looks after her. (3.1)

Sentence (3.1) has a definite description (*her husband*) as part of the consequent of an implication, with the antecedent setting up the requirements for uniqueness of reference of that description (the introduction of *a woman*, together with the assertion that she is married). The linking of the antecedent *a woman* with the possessive pronoun *her* and with the personal pronoun *her* are cases of so-called unselective binding. Unselective binding is essentially a dynamic phenomenon; the traditional 'static' approaches to NL meaning representation (e.g., in the style of Montague grammar) cannot deal with it in a convincing way.

In this chapter we focus on a logical analysis of DPL as an NL knowledge representation medium. The structure of the chapter is a follows. In Section 3.2 we review the semantics of dynamic predicate logic, i.e., the system of Groenendijk and Stokhof [8], extended with a clause for definite assignment. To axiomatize this logic, we propose the medium of Quantified Dynamic Logic, which is presented in Section 3.3. Section 3.4 contains a digression on the concept of dynamic entailment in dynamic predicate logic. In Section 3.5 a Pratt-style axiomatisation for proper state dynamic semantics is presented. The importance of the enterprise of finding a calculus for DPL will not be stressed here, but the issue is discussed at length in Van Eijck and De Vries [4]. The calculus of the present chapter is more general than the Hoare style calculus of [4]: it has greater expressive power, and the rules from [4] can all be derived in the new calculus (Section 3.6). Section 3.7 demonstrates the use of the calculus by giving an example derivation of the static meaning of the DPL program for sentence (3.1).

3.2 Dynamic Predicate Logic and Its Semantics

In this section we define DPL, the language of dynamic predicate logic with ι assignment, and its semantics. Let a set of variables V and a set of constants C plus a set of relation symbols with their arities (a signature L) be given. Then $V \cup C$ are the terms of DPL_L, and DPL_L itself, the set of DPL programs over signature L, is defined as follows.

DEFINITION 3.1 (SYNTAX OF DPL_L) DPL_L is the smallest set for which the following hold:

1. If t_1, t_2 are terms, then $t_1 = t_2$ is a DPL program.

2. If R is an n-place relation symbol and $t_1, \ldots, t_n$ are terms, then $R(t_1 \cdots t_n)$ is a DPL program.

3. If π_1 and π_2 are DPL programs then $(\pi_1; \pi_2)$ is a DPL program.

4. If π_1 and π_2 are DPL programs then $(\pi_1 \Rightarrow \pi_2)$ is a DPL program.

5. If π is a program, then $\neg\pi$ is a DPL program.

6. If π is a DPL program and x is a variable, then $\eta x : \pi$ is a DPL program.

7. If π is a DPL program and x is a variable, then $\iota x : \pi$ is a DPL program.

We will follow the usual predicate logical convention of omitting outermost parentheses for readability. Also, it will become evident from the semantic clause for sequential composition that the ; operator is associative. Therefore, we will often take the liberty to write $\pi_1; \pi_2; \pi_3$ instead of $(\pi_1; \pi_2); \pi_3$ or $\pi_1; (\pi_2; \pi_3)$. Also, $t_1 \neq t_2$ will be used as an abbreviation of $\neg t_1 = t_2$, $\top$ as an abbreviation for $\neg\eta v_0 : v_0 \neq v_0$, and $\bot$ as an abbreviation for $\neg\top$.

Given a model $\mathcal{M} = \langle M, F \rangle$, with M a universe of individuals and F an interpretation for the signature of the language, a proper state for $\mathcal{M}$ is a function in M^V. We will refer to the set of proper states for $\mathcal{M}$ as $\mathbf{S}_{\mathcal{M}}$.

We will spell out the formal details of the standard semantics for dynamic predicate logic with ι assignment. This semantics is called *proper state semantics* to emphasise that the definition is couched in terms of proper states only. A semantics for DPL that allows for the possibility of error abortion of a program to take presuppositions of program execution into account (error state semantics) is proposed in [2, 3].

A proper state s for $\mathcal{M} = \langle M, F \rangle$ determines a valuation $\mathbf{V}_s$ for the terms of the language as follows: if $t \in V$ then $\mathbf{V}_s(t) = s(t)$, if $t \in C$, then $\mathbf{V}_s(t) = F(c)$. The valuation function $\mathbf{V}_s$ is used in the definition of the notion of satisfaction for atomic formulae.

DEFINITION 3.2

- $\mathcal{M} \models_s Rt_1 \cdots t_n$ if $\langle \mathbf{V}_s(t_1), \ldots, \mathbf{V}_s(t_n) \rangle \in F(R)$,
- $\mathcal{M} \models_s t_1 = t_2$ if $\mathbf{V}_s(t_1) = \mathbf{V}_s(t_2)$.

This concept will play a role in the semantic clauses for atomic programs.

If s is a proper state for $\mathcal{M}$, v a variable and d an element of the universe or $\mathcal{M}$, then $s(v|d)$ is the proper state for $\mathcal{M}$ which is just like s except for the possible difference that v is mapped to d.

We define a function $[\![\pi]\!]_{\mathcal{M}} : \mathbf{S}_{\mathcal{M}} \to \mathcal{P}\mathbf{S}_{\mathcal{M}}$ by recursion. s, t are used as metavariables over (proper) states. The function $[\![\pi]\!]_{\mathcal{M}}$ depends on the model $\mathcal{M}$, but for convenience I will often write $[\![\pi]\!]$ rather than $[\![\pi]\!]_{\mathcal{M}}$. The function should be read as: on input state s, π may produce any of the outputs in output state set $[\![\pi]\!](s)$.

DEFINITION 3.3 (SEMANTICS OF DPL)

1. $[\![Rt_1 \cdots t_n]\!](s) = \begin{cases} \{s\} & \text{if } \mathcal{M} \models_s Rt_1 \cdots t_n \, , \\ \emptyset & \text{otherwise.} \end{cases}$

2. $[\![t_1 = t_2]\!](s) = \begin{cases} \{s\} & \text{if } \mathcal{M} \models_s t_1 = t_2 \, , \\ \emptyset & \text{otherwise.} \end{cases}$

3. $[\![(\pi_1; \pi_2)]\!](s) = \bigcup \{ [\![\pi_2]\!](u) \mid u \in [\![\pi_1]\!](s) \}$.

4. $[\![(\pi_1 \Rightarrow \pi_2)]\!](s) = \begin{cases} \{s\} & \text{if for all } u \in [\![\pi_1]\!](s) \text{ it holds that} \\ & [\![\pi_2]\!](u) \neq \emptyset, \\ \emptyset & \text{otherwise.} \end{cases}$

5. $[\![\neg\pi]\!](s) = \begin{cases} \{s\} & \text{if } [\![\pi]\!](s) = \emptyset, \\ \emptyset & \text{otherwise.} \end{cases}$

6. $[\![\eta x : \pi]\!](s) = \bigcup \{ [\![\pi]\!](s(x|d)) \mid d \in U \}$.

7. $[\![\iota x : \pi]\!](s) = \begin{cases} [\![\pi]\!](s(x|d)) \text{ for the unique } d \in U \text{ s.t. } [\![\pi]\!](s(x|d)) \neq \emptyset & \text{if } d \text{ exists,} \\ \emptyset & \text{otherwise.} \end{cases}$

The statement $\eta x : \pi$ performs a non-deterministic action, for it sanctions any assignment to x of an individual satisfying π. The statement acts as a test at the same time: in case there are no individuals satisfying π the set of output states for any given input state will be empty. In fact, the meaning of $\eta x : \pi$ can be thought of as a random or non-deterministic assignment followed by a test, for $\eta x : \pi$ is equivalent to $\eta x : \top; \pi$, or in more standard notation, $x := ?; \pi$. It follows immediately from this explanation plus the dynamic meaning of sequential composition that $\eta x : (\pi_1); \pi_2$ is equivalent to $\eta x : (\pi_1; \pi_2)$.

The interpretation conditions for ι assignment make clear how the uniqueness condition is handled dynamically. The statement $\iota x : \pi$ consists of a test followed by a deterministic action in case the test succeeds: first it is checked whether there is a unique d for which $\pi(x)$ succeeds; if so, this individual is assigned to x and π is performed; otherwise the program fails (in other words, the set of output states is empty). It is not difficult to see that this results in the Russell treatment for definite descriptions [12]. Also, we see that the two programs $\iota x : (\pi_1); \pi_2$ and $\iota x : (\pi_1; \pi_2)$ are not equivalent, i.e., that there are states on which they have different effects. The program $\iota x : (\pi_1; \pi_2)$ succeeds if there is a unique object d satisfying $\pi_1; \pi_2$, while the requirement for $\iota x : (\pi_1); \pi_2$ is stronger: there has to be a unique individual d satisfying π_1, and d must also satisfy π_2.

3.3 Quantified Dynamic Logic

Our aim in this chapter is to supplement the proper state semantics of our representation language with an axiom system in the style of Pratt [11]. We will also indicate how the calculus relates to the Hoare style axiomatisation (based on rules in the spirit of [1, 9]) for DPL that was proposed in [4].

A Hoare style axiomatisation of proper state semantics for DPL uses so-called universal and existential correctness statements that relate static assertions from predicate logic to pre- and postconditions of DPL programs. These Hoare correctness statements have the form of implications: if precondition φ holds of a state, then in all output states of program π for that input, statement ψ will hold (universal correctness), or: if precondition φ holds of a state, then in at least one output states of program π for that input, statement ψ will hold (existential correctness).

If we want to be able to use the full range of logical connections between static assertions from predicate logic and programs from DPL we need a more powerful representation medium. We will define a version of quantified dynamic logic (inspired by Pratt's dynamic logic [7, 10, 11]), that gives us the expressive power we need. Let L be the same signature (set of relations, with their arities, plus individual constants) that was used in the definition of DPL. Let V be the same countable set of variables that was used in the definition of DPL and C the set of constants in the signature L. V and C together form the terms of qd_L.

DEFINITION 3.4 (SYNTAX OF qd_L) $form(qd_L)$ and $prog(qd_L)$ are the smallest sets for which the following hold:

1. If R is an n-place relation in L, and $t_1, \ldots, t_n$ are terms, then $Rt_1 \cdots t_n \in form(qd_L)$.

2. If t_1, t_2 are terms, then $t_1 = t_2 \in form(qd_L)$.

3. If $\varphi, \psi \in form(qd_L)$, then $(\varphi \wedge \psi), \neg\varphi \in form(qd_L)$.

4. If $v \in V$ and $\varphi \in form(qd_L)$, then $\exists v\varphi \in form(qd_L)$.

5. If $\pi \in prog(qd_L)$ and $\varphi \in form(qd_L)$, then $\langle \pi \rangle \varphi \in form(qd_L)$.

6. If R is an n-place relation in L, and $t_1, \ldots, t_n$ are terms, then $\boldsymbol{Rt_1 \cdots t_n}, \boldsymbol{t_1 = t_2} \in prog(qd_L)$.

7. If $\boldsymbol{\pi}, \boldsymbol{\pi'} \in prog(qd_L)$ then $\neg\boldsymbol{\pi}, \boldsymbol{\pi}; \boldsymbol{\pi'}, \boldsymbol{\pi} \Rightarrow \boldsymbol{\pi'} \in prog(qd_L)$.

8. If $v \in V$ and $\boldsymbol{\pi} \in prog(qd_L)$, then $\boldsymbol{\eta v : \pi}, \boldsymbol{\iota v : \pi} \in prog(qd_L)$.

Note that the programs of qd_L are the programs of DPL_L. We use boldface for the test program $\boldsymbol{Rt_1 \cdots t_n}$ and italics for the formula $Rt_1 \cdots t_n$. We will continue to use the abbreviation conventions with respect to DPL programs in the qd_L format. As is customary, we abbreviate $\neg(\neg\varphi \wedge \neg\psi)$ as $(\varphi \vee \psi)$, $\neg(\varphi \wedge \neg\psi)$ as $(\varphi \rightarrow \psi)$, $(\varphi \rightarrow \psi) \wedge (\psi \rightarrow \varphi)$ as $\varphi \leftrightarrow \psi$, $\neg\langle \pi \rangle \neg\varphi$ as $[\pi]\varphi$ and $\neg\exists x\neg\varphi$ as $\forall x\varphi$. Also, we omit outermost parentheses for readability. Finally, we add the convention for qd_L formulae that $\top$ is an abbreviation of $\forall x(x = x)$ and $\bot$ an abbreviation of $\neg\top$.

The semantics of qd_L is given in terms of natural models for qd_L, where a *natural model* consists of a first order model $\mathcal{M}$ for the signature L, and a set of two place relations R_π on M^V, one for each program in $prog(qd_L)$, with the programs interpreted as given by the semantics of DPL, i.e., $\langle s, t \rangle \in R_\pi$ iff $t \in [\![\pi]\!]_{\mathcal{M}}(s)$. As a natural model is fully determined by its 'static' part $\mathcal{M}$, we will continue to refer to the natural model based on $\mathcal{M}$ as $\mathcal{M}$. For further observations on the relations between first order models and the natural models based on them see Fernando [6].

We can now successively define the notions of *satisfaction* of a qd_L formula by a state s for a model $\mathcal{M}$, of *truth* of a qd_L formula for a model $\mathcal{M}$, of *(universal) validity* of a qd_L formula, and finally of *local and global consequence* for a class of qd_L formulae and a qd_L formula.

DEFINITION 3.5 (SATISFACTION FOR qd_L)

1. $\mathcal{M} \models_s Rt_1 \cdots t_n$ is given by the standard Tarski satisfaction definition.

2. $\mathcal{M} \models_s t_1 = t_2$ is again given by the standard Tarski satisfaction definition.

3. $\mathcal{M} \models_s \neg\varphi$ if it is not the case that $\mathcal{M} \models_s \varphi$.

4. $\mathcal{M} \models_s \varphi \wedge \psi$ if $\mathcal{M} \models_s \varphi$ and $\mathcal{M} \models_s \psi$.

5. $\mathcal{M} \models_s \exists v\varphi$ if for some $d \in M$, $\mathcal{M} \models_{s(v|d)} \varphi$.

6. $\mathcal{M} \models_s \langle \pi \rangle \varphi$ if there is some $t \in [\![\pi]\!]_{\mathcal{M}}(s)$ with $\mathcal{M} \models_t \varphi$.

DEFINITION 3.6 (TRUTH FOR qd_L) $\mathcal{M} \models \varphi$ if for all $s \in M^V$: $\mathcal{M} \models_s \varphi$.

DEFINITION 3.7 (VALIDITY FOR qd_L) $\models \varphi$ if for all $\mathcal{M}$: $\mathcal{M} \models \varphi$.

DEFINITION 3.8 ('LOCAL' CONSEQUENCE FOR qd_L) $\Gamma \models \varphi$ if for all pairs $\mathcal{M}, s$ with $s \in M^V$ and $\mathcal{M} \models_s \gamma$ for every $\gamma \in \Gamma$, it holds that $\mathcal{M} \models_s \varphi$.

This consequence notion is called 'local' because it is phrased in terms of satisfaction for model/state pairs, where a 'global' definition is phrased in terms of truth:

DEFINITION 3.9 ('GLOBAL' CONSEQUENCE FOR qd_L) $\Gamma \models^* \varphi$ if for all $\mathcal{M}$ with $\mathcal{M} \models \gamma$ for every $\gamma \in \Gamma$, it holds that $\mathcal{M} \models \varphi$.

PROPOSITION 3.1 $\Gamma \models \varphi$ entails $\Gamma \models^* \varphi$.

Proof Immediate from the definitions.   ∎

In fact, the local notion is the more fine-grained one. Writing $\bigwedge \Gamma$ for the result of taking the universal closures of all the formulae in Γ (where the universal closure of a qd_L formula also closes off the variables in the test predicates, e.g., the universal closure of $\langle Rxy \rangle \top$ is $\forall x \forall y \langle Rxy \rangle \top$), we can state the relation between the two notions as follows:

PROPOSITION 3.2 $\Gamma \models^* \varphi$ iff $\bigwedge \Gamma \models \varphi$.

Proof Immediate from the fact that $\mathcal{M} \models \forall \bar{x} \gamma(\bar{x})$ iff for all $s \in M^V$ it holds that $\mathcal{M} \models_s \gamma(\bar{x})$.   ∎

The distinction between local and global consequence also applies to modal logic: Venema [13] makes a solid case for local consequence in modal logic and discusses the differences between the matching deduction relations $\vdash$ and $\vdash^*$ that the distinction engenders.

This completes the exposition of the tools we need for axiomatizing the proper state semantics from Section 3.2.

Before we look at deduction for quantified dynamic logic we will briefly digress to a discussion of various possible entailment relations for programs.

3.4   Dynamic Entailment for DPL Programs

One of the key notions that proponents of dynamic interpretation are trying to capture is the notion of dynamic entailment. One very obvious choice is to let program π_1 dynamically entail program π_2 if the weakest precondition for success of the first program logically entails the weakest precondition for success of the second. Formally:

DEFINITION 3.10 π_1 dynamically entails$_1$ π_2 (notation $\pi_1 \models_1 \pi_2$) if $\mathcal{M} \models_s \langle \pi_1 \rangle \top$ implies $\mathcal{M} \models_s \langle \pi_2 \rangle \top$.

Here is an observation made by Johan van Benthem (personal communication) on entailment relation $\models_1$.

PROPOSITION 3.3 $\models_1$ satisfies the following structural rules:

- Reflexivity: $\dfrac{}{\pi \Longrightarrow \pi}$.

- Right monotonicity: $\dfrac{\pi_1; \cdots; \pi_n \Longrightarrow \pi_c}{\pi_1; \cdots; \pi_n; \pi \Longrightarrow \pi_c}$

- Right cut: $\dfrac{\pi_1; \cdots; \pi_n; \pi \Longrightarrow \pi_c \qquad \pi_{n+1}; \cdots; \pi_m \Longrightarrow \pi}{\pi_1; \cdots \pi_n; \pi_{n+1}; \cdots; \pi_m \Longrightarrow \pi_c}$.

Conversely, every dynamic consequence relation $\Longrightarrow$ satisfying these three rules can be represented in dynamic terms as follows: the domain of the entailing program is contained in the domain of the entailed program.

Proof   It is easily checked that $\models_1$ satisfies reflexivity, right monotonicity and right cut.

Conversely, let, for some language L, a relation $\Longrightarrow$ satisfying reflexivity, right monotonicity and right cut be given. Take as states the finite sequences of propositions from L, take as relations the sets of state pairs $[\![P]\!] = \{\langle XY, Y \rangle \mid X \Longrightarrow P\}$, and check the statement that $P_1; \cdots; P_n \Longrightarrow C$ iff $\mathrm{dom}([\![P_1]\!] \circ \cdots \circ [\![P_n]\!]) \subseteq \mathrm{dom}[\![C]\!]$.   ∎

The relation $\models_1$ includes a smaller relation $\models_2$ of 'meaning inclusion', which can be characterized in the present framework as follows.

DEFINITION 3.11 $\pi_1 \models_2 \pi_2$ if for all Predicate Logical φ (formulae without occurrences of program modalities), for all $\mathcal{M}$, for all $s \in M^V$, $\mathcal{M} \models_s \langle \pi_1 \rangle \varphi$ implies $\mathcal{M} \models_s \langle \pi_2 \rangle \varphi$.

Immediately from the definitions we get:

PROPOSITION 3.4   If $\pi_1 \models_2 \pi_2$ then $\pi_1 \models_1 \pi_2$.

Several other options for dynamic entailment present themselves. The choice made by Groenendijk and Stokhof in [8] takes the following shape in our framework.

DEFINITION 3.12 π_1 dynamically entails$_3$ π_2 (notation $\pi_1 \models_3 \pi_2$) if $\models [\pi_1]\langle \pi_2 \rangle \top$.

The main motivation for this choice of entailment notion is the following proposition.

PROPOSITION 3.5 $\pi_1 \models_3 \pi_2$ iff $\models \langle \pi_1 \Rightarrow \pi_2 \rangle \top$.

Proof $\pi \models_3 \pi_2$ iff $\models [\pi_1]\langle \pi_2 \rangle \top$
iff for all $\mathcal{M}$, all $s \in M^V$, $\mathcal{M} \models_s [\pi_1]\langle \pi_2 \rangle \top$
iff for all $\mathcal{M}$, all $s \in M^V$, for all $t \in [\![\pi_1]\!]_{\mathcal{M}}(s)$, $\mathcal{M} \models_t \langle \pi_2 \rangle \top$
iff for all $\mathcal{M}$, all $s \in M^V$, for all $t \in [\![\pi_1]\!]_{\mathcal{M}}(s)$, $[\![\pi_2]\!]_{\mathcal{M}}(t) \neq \emptyset$
iff for all $\mathcal{M}$, all $s \in M^V$, $\mathcal{M} \models_s \langle \pi_1 \Rightarrow \pi_2 \rangle \top$
iff $\models \langle \pi_1 \Rightarrow \pi_2 \rangle \top$.   ∎

It would be worth one's while to further investigate these and other possible choices for the dynamic entailment relation. Here is a further example of an entailment notion which, at least to the present author, looks every bit as 'natural' as the earlier ones.

DEFINITION 3.13 π_1 dynamically entails$_4$ π_2 (notation $\pi_1 \models_4 \pi_2$) if $\mathcal{M} \models_s \langle \pi_1 \rangle \top$ implies $\mathcal{M} \models_s \langle \pi_1 \rangle \langle \pi_2 \rangle \top$.

PROPOSITION 3.6 $\pi_1 \models_4 \pi_2$ iff $\models \langle \neg\neg\pi_1 \Rightarrow (\pi_1; \pi_2) \rangle \top$.

Proof $\pi \models_4 \pi_2$ iff for all $\mathcal{M}$, all $s \in M^V$, $\mathcal{M} \models_s \langle \pi_1 \rangle \top$ implies $\mathcal{M} \models_s \langle \pi_1 \rangle \langle \pi_2 \rangle \top$
iff for all $\mathcal{M}$, all $s \in M^V$, $[\![\neg\neg\pi_1]\!]_{\mathcal{M}}(s) = \{s\}$ implies $[\![\pi_1; \pi_2]\!]_{\mathcal{M}}(s) \neq \emptyset$
iff for all $\mathcal{M}$, all $s \in M^V$, $\mathcal{M} \models_s \langle \neg\neg\pi_1 \Rightarrow (\pi_1; \pi_2) \rangle \top$
iff $\models \langle \neg\neg\pi_1 \Rightarrow (\pi_1; \pi_2) \rangle \top$.   ∎

The notions $\models_1, \models_2, \models_3$ and $\models_4$ are couched in terms of $\models_s$ (they are exemplifications of the local perspective on entailment), and of course there are variants where the universal quantification over states is distributed (exemplifications of the global perspective on entailment).

DEFINITION 3.14

- π_1 dynamically entails$_{1*}$ π_2 (notation $\pi_1 \models_{1*} \pi_2$) if $\models \langle \pi_1 \rangle \top$ implies $\models \langle \pi_2 \rangle \top$.
- π_1 dynamically entails$_{2*}$ π_2 (notation $\pi_1 \models_{2*} \pi_2$) if $\models \langle \pi_1 \rangle \varphi$ implies $\models \langle \pi_2 \rangle \varphi$ (for PL φ).
- π_1 dynamically entails$_{3*}$ π_2 (notation $\pi_1 \models_{3*} \pi_2$) if $\models \langle \pi_1 \rangle \top$ implies $\models [\pi_1] \langle \pi_2 \rangle \top$.
- π_1 dynamically entails$_{4*}$ π_2 (notation $\pi_1 \models_{4*} \pi_2$) if $\models \langle \pi_1 \rangle \top$ implies $\models \langle \pi_1 \rangle \langle \pi_2 \rangle \top$.

The following proposition is immediate from the definitions.

PROPOSITION 3.7

1. $\pi_1 \models_1 \pi_2$ entails $\pi_1 \models_{1*} \pi_2$.
2. $\pi_1 \models_2 \pi_2$ entails $\pi_1 \models_{2*} \pi_2$.
3. $\pi_1 \models_3 \pi_2$ entails $\pi_1 \models_{3*} \pi_2$.
4. $\pi_1 \models_4 \pi_2$ entails $\pi_1 \models_{4*} \pi_2$.

Again, the local perspective on dynamic consequence turns out to be the more fine grained one. For our next proposition we need the universal closure of a program, i.e., the universal dynamic quantification $\neg \eta x \neg$ over all variables with free occurrences in the program. We define fv, the function which gives the set of variables with free occurrences in a program, and av, the function which gives the set of variables with active occurrences in the program, by simultaneous recursion.

DEFINITION 3.15

$$fv(Rt_1 \cdots t_n) = \{t_i \mid t_i \text{ is a variable}\}, \qquad av(Rt_1 \cdots t_n) = \emptyset.$$
$$fv(t_1 = t_2) = \{t_i \mid t_i \text{ is a variable}\}, \qquad av(t_1 = t_2) = \emptyset.$$
$$fv(\neg \pi) = fv(\pi), \qquad av(\pi) = \emptyset.$$
$$fv(\pi_1; \pi_2) = fv(\pi_1) \cup (fv(\pi_2) - av(\pi_1)), \qquad av(\pi_1; \pi_2) = av(\pi_1) \cup av(\pi_2).$$
$$fv(\pi_1 \Rightarrow \pi_2) = fv(\pi_1) \cup (fv(\pi_2) - av(\pi_1)), \qquad av(\pi_1 \Rightarrow \pi_2) = \emptyset.$$
$$fv(\eta x : \pi) = fv(\pi) - \{x\}, \qquad av(\eta x : \pi) = av(\pi_1) \cup \{x\}.$$
$$fv(\iota x : \pi) = fv(\pi) - \{x\}, \qquad av(\iota x : \pi) = av(\pi_1) \cup \{x\}.$$

Writing $\neg \eta \bar{x} \neg \pi(\bar{x})$ for $\neg \eta x_1 : \cdots \eta x_n : \neg \pi$, where $\{x_1, \ldots, x_n\} = fv(\pi)$, we can state the following proposition.

PROPOSITION 3.8

1. $\neg \eta \bar{x} \neg \pi_1(\bar{x}) \models_1 \pi_2$ iff $\pi_1 \models_{1*} \pi_2$.
2. $\neg \eta \bar{x} \neg \pi_1(\bar{x}) \models_2 \pi_2$ iff $\pi_1 \models_{2*} \pi_2$.

3. $\neg \eta \bar{x} \neg \pi_1(\bar{x}) \models_3 \pi_2$ iff $\pi_1 \models_{3*} \pi_2$.

4. $\neg \eta \bar{x} \neg \pi_1(\bar{x}) \models_4 \pi_2$ iff $\pi_1 \models_{4*} \pi_2$.

Proof For 1, 3, 4, the result follows from the fact that $\mathcal{M} \models \langle \neg \eta \bar{x} \neg \pi(\bar{x}) \rangle \top$ iff for all $s \in M^V$ it holds that $\mathcal{M} \models_s \langle \pi \rangle \top$, and for 2 it follows from the fact that $\mathcal{M} \models \langle \neg \eta \bar{x} \neg \pi(\bar{x}) \rangle \varphi$ iff for all $s \in M^V$ it holds that $\mathcal{M} \models_s \langle \pi \rangle \varphi$ (for φ PL). ∎

3.5 A Calculus for QDL with Proper State Semantics

The calculus for the semantics of quantified dynamic logic based on proper state semantics for dynamic predicate logic has four sets of axiom schemata: (i) propositional and quantificational schemata, (ii) K-schemata, (iii) atomic test schemata, and (iv) program composition schemata.

Propositional and Quantificational Schemata We start by taking the axiom schemata of propositional logic and first order quantification:

A 1 $\varphi \rightarrow (\psi \rightarrow \varphi)$.

A 2 $(\varphi \rightarrow (\psi \rightarrow \chi)) \rightarrow ((\varphi \rightarrow \psi) \rightarrow (\varphi \rightarrow \chi))$.

A 3 $(\neg \varphi \rightarrow \neg \psi) \rightarrow (\psi \rightarrow \varphi)$.

A 4 $\forall v \varphi \rightarrow [t/v]\varphi$, provided t is free for v in φ.

A 5 $\varphi \rightarrow \forall v \varphi$, provided v has no free occurrences in φ.

A 6 $\forall v(\varphi \rightarrow \psi) \rightarrow (\forall v \varphi \rightarrow \forall v \psi)$.

A 7 $v = v$.

A 8 $v = w \rightarrow (\varphi \rightarrow \varphi')$, where φ' results from replacing some v-occurrence(s) in φ by w.

See, e.g., Enderton [5] for discussion and motivation.

Atomic Test Schemata

A 9 $\langle Rt_1 \cdots t_n \rangle \varphi \leftrightarrow (Rt_1 \cdots t_n \wedge \varphi)$.

A 10 $\langle t_1 = t_2 \rangle \varphi \leftrightarrow (t_1 = t_2 \wedge \varphi)$.

Program Composition Schemata   The schemata for complex programs.

A 11 $\langle \pi_1; \pi_2 \rangle \varphi \leftrightarrow \langle \pi_1 \rangle \langle \pi_2 \rangle \varphi$.

A 12 $\langle \neg \pi \rangle \varphi \leftrightarrow (\varphi \wedge [\pi] \bot)$.

A 13 $\langle \pi_1 \Rightarrow \pi_2 \rangle \varphi \leftrightarrow (\varphi \wedge [\pi_1] \langle \pi_2 \rangle \top)$.

A 14 $\langle \eta x : \pi \rangle \varphi \leftrightarrow \exists x \langle \pi \rangle \varphi$.

A 15 $\langle \iota x : \pi \rangle \varphi \leftrightarrow (\exists! x \langle \pi \rangle \top \wedge \exists x \langle \pi \rangle \varphi)$.

Rules of Inference   The rules of inference are as follows.

R 1 (Universal Generalization)   Conclude from $\vdash \varphi$ to $\vdash \forall v \varphi$.

R 2 (Necessitation)   For every program modality $\langle \pi \rangle$: conclude from $\vdash \varphi$ to $\vdash [\pi] \varphi$.

R 3 (Modus Ponens)   Conclude from $\vdash \varphi \to \psi$ and $\vdash \varphi$ to $\vdash \psi$.

The notion of theoremhood in the calculus is standard.

DEFINITION 3.16   Formula φ is a theorem of the calculus, notation $\vdash \varphi$, if φ fits one of the axiom schemata or φ follows from theorems in the calculus by an application of one of the inference rules.

In formulating a notion of deducibility we have to bear in mind that the notion is intended to match the local notion $\models$ for consequence in QDL. We must make sure that $\Gamma \vdash \varphi$ means that the hypotheses from Γ are interpreted in such a way that they set up a context for the free variables occurring in them. This contrasts with a setup where the formulae from Γ are being handled as extra axioms (admitting universal generalization).

DEFINITION 3.17   Formula φ is deducible in the calculus from formula collection Γ, notation $\Gamma \vdash \varphi$, if there are $\gamma_1, \ldots, \gamma_n \in \Gamma$, with $n \geq 0$, such that $\vdash (\gamma_1 \wedge \cdots \wedge \gamma_n) \to \varphi$.

Note that this local notion of derivability is more fine-grained than the global notion (that we will not bother to spell out). Indeed, the relation between the two notions of derivability is intended to mirror the relation between the local and the global notions of consequence (Proposition 3.2).

As is customary, we write $\varphi \vdash \psi$ for $\{\varphi\} \vdash \psi$. The fact that $\vdash$ is a local notion entails that we get the deduction theorem for free.

PROPOSITION 3.9 (DEDUCTION) $\varphi \vdash \psi$ iff $\vdash \varphi \to \psi$.

Proof Immediate from Definition 3.17.   ∎

PROPOSITION 3.10 (SUBSTITUTION)   If $\Gamma \vdash \varphi$ and $\Gamma \vdash \psi \leftrightarrow \chi$, and φ' is the result of substituting occurrences of χ for occurrences of ψ in φ, then $\Gamma \vdash \varphi'$.

Proof Induction on the structure of φ.   ∎

THEOREM 3.1   The K-schema is derivable for every program π:
$\vdash [\pi](\varphi \to \psi) \to ([\pi]\varphi \to [\pi]\psi)$.

Proof Induction on the structure of π.   ∎

THEOREM 3.2   The calculus is sound for natural models, i.e., if $\Gamma \vdash \varphi$ then $\Gamma \models \varphi$.

Proof Standard checking of the axioms and rules.   ∎

THEOREM 3.3   The calculus is complete for natural models, i.e., if $\Gamma \models \varphi$ then $\Gamma \vdash \varphi$.

Proof First observe that the following translation function * from qd_L to first order predicate logic over L preserves satisfaction.

$$
\begin{aligned}
(Rt_1 \cdots t_n)^* &= Rt_1 \cdots t_n \\
(t_1 = t_2)^* &= t_1 = t_2 \\
(\varphi \wedge \psi)^* &= \varphi^* \wedge \psi^* \\
(\neg\varphi)^* &= \neg\varphi^* \\
(\exists v\varphi)^* &= \exists v\varphi^* \\
(\langle Rt_1 \cdots t_n \rangle \varphi)^* &= \varphi^* \wedge Rt_1 \cdots t_n \\
(\langle t_1 = t_2 \rangle \varphi)^* &= \varphi^* \wedge t_1 = t_2 \\
(\langle \pi_1; \pi_2 \rangle \varphi)^* &= (\langle \pi_1 \rangle \langle \pi_2 \rangle \varphi)^* \\
(\langle \neg\pi \rangle \varphi)^* &= \varphi^* \wedge ([\pi]\bot)^* \\
(\langle \eta v : \pi \rangle \varphi)^* &= \exists v(\langle \pi \rangle \varphi)^*
\end{aligned}
$$

Thus, it follows from $\Gamma \models \varphi$ that $\Gamma \models \varphi^*$. Next, use the completeness of first order predicate logic to conclude from $\Gamma \models \varphi^*$ that $\Gamma \vdash \varphi^*$. Finally, note that the translation steps and their inverses in the definition of * are licenced by the atomic test schemata and the program composition schemata of the calculus. This allows us to conclude from $\Gamma \vdash \varphi^*$ that $\Gamma \vdash \varphi$.   ∎

3.6   Hoare Style Rules as Derived Rules of the Calculus

It is instructive to see how the Hoare style rules for DPL of Van Eijck and De Vries [4] arise as derived rules of inference in the present calculus.

The test axioms from [4] are the two sides of the bi-implication given by the test axiom (9). Here is the derivation of the existential test axiom.

1. $\langle \boldsymbol{Rt_1} \cdots t_n \rangle \varphi \leftrightarrow (Rt_1 \cdots t_n \wedge \varphi).$ axiom (9)
2. $(\langle \boldsymbol{Rt_1} \cdots t_n \rangle \varphi \leftrightarrow (Rt_1 \cdots t_n \wedge \varphi))$
    $\rightarrow ((Rt_1 \cdots t_n \wedge \varphi) \rightarrow \langle \boldsymbol{Rt_1} \cdots t_n \rangle \varphi)$ PL
3. $(Rt_1 \cdots t_n \wedge \varphi) \rightarrow \langle \boldsymbol{Rt_1} \cdots t_n \rangle \varphi.$ 1,2,MP.

Here is the derivation of the universal test axiom.

1. $\langle \boldsymbol{Rt_1} \cdots t_n \rangle \varphi \leftrightarrow (Rt_1 \cdots t_n \wedge \varphi).$ axiom (9)
2. $(\langle \boldsymbol{Rt_1} \cdots t_n \rangle \varphi \leftrightarrow (Rt_1 \cdots t_n \wedge \varphi))$
    $\rightarrow ((Rt_1 \cdots t_n \rightarrow \varphi) \rightarrow [\boldsymbol{Rt_1} \cdots t_n] \varphi)$ PL+ def $[\pi]$
3. $(Rt_1 \cdots t_n \rightarrow \varphi) \rightarrow [\boldsymbol{Rt_1} \cdots t_n] \varphi$ 1, 2, MP.

The effects of the oracle rule of [4], stating that all PL theorems are axioms, are taken care of in the present setup by the propositional and quantificational axioms.

The consequence rules for existential and universal correctness of [4] each can be established as derived consequence rules in the present calculus. Here is the existential consequence rule:

$$\frac{\varphi \rightarrow \psi \qquad \psi \rightarrow \langle \pi \rangle \chi \qquad \chi \rightarrow \xi}{\varphi \rightarrow \langle \pi \rangle \xi.}$$

Here is its derivation:

1. φ hypothesis
2. $\varphi \rightarrow \psi$ premiss
3. ψ 1, 2, MP
4. $\psi \rightarrow \langle \pi \rangle \chi$ premiss
5. $\langle \pi \rangle \chi$ 3, 4, MP
6. $\chi \rightarrow \xi$ premiss
7. $\langle \pi \rangle \chi \rightarrow \langle \pi \rangle \xi$ 6, Nec, K-axiom
8. $\langle \pi \rangle \xi$ 5, 7, MP
9. $\varphi \rightarrow \langle \pi \rangle \xi$ 1, 8, DED.

Here is the universal consequence rule:

$$\frac{\varphi \to \psi \qquad \psi \to [\pi]\chi \qquad \chi \to \xi}{\varphi \to \langle\pi\rangle\xi.}$$

Here is its derivation:

1.	φ	hypothesis
2.	$\varphi \to \psi$	premiss
3.	ψ	1, 2, MP
4.	$\psi \to [\pi]\chi$	premiss
5.	$[\pi]\chi$	3, 4, MP
6.	$\chi \to \xi$	premiss
7.	$[\pi]\chi \to [\pi]\xi$	6, Nec, K-axiom
8.	$[\pi]\xi$	5, 7, MP
9.	$\varphi \to [\pi]\xi$	1, 8, DED.

Similarly, all other rules from [4] can be derived in the calculus.

As was remarked earlier, the premisses and conclusion in a Hoare style rule give preconditions which are not necessarily weakest preconditions, or equivalently, postconditions which are not necessarily strongest postconditions. For purposes of reasoning about DPL programs, however, derived rules yielding conclusions giving weakest preconditions of complex programs from premisses giving information about weakest preconditions of their components are much more useful. Note that such rules cannot be stated in the Hoare format, as its premisses and conclusion do not have the form of implications.

We will now derive a number of rules for calculating weakest preconditions of complex programs in terms of the weakest preconditions of their component programs.

Here is a derived rule for sequential composition.

$$\frac{\varphi \leftrightarrow \langle\pi_1\rangle\psi \qquad \psi \leftrightarrow \langle\pi_2\rangle\chi}{\varphi \leftrightarrow \langle\pi_1; \pi_2\rangle\chi.}$$

And here is its derivation:

1. $\varphi \leftrightarrow \langle\pi_1\rangle\psi$ premiss
2. $\psi \leftrightarrow \langle\pi_2\rangle\chi$ premiss
3. $\varphi \leftrightarrow \langle\pi_1\rangle\langle\pi_2\rangle\chi$ 1, 2, SUBST
4. $\varphi \leftrightarrow \langle\pi_1; \pi_2\rangle\chi$ 3, ;-axiom, SUBST.

A dual to this derived rule for sequential composition is also easily derived.

$$\frac{\varphi \leftrightarrow [\pi_1]\psi \qquad \psi \leftrightarrow [\pi_2]\chi}{\varphi \leftrightarrow [\pi_1; \pi_2]\chi.}$$

A derived rule of negation:

$$\frac{\varphi \leftrightarrow [\pi]\bot}{(\varphi \wedge \psi) \leftrightarrow \langle \neg \pi \rangle \psi.}$$

Its derivation:

1. $\varphi \leftrightarrow [\pi]\bot$ premiss
2. $([\pi]\bot \wedge \psi) \leftrightarrow \langle \neg \pi \rangle \psi$ $\neg$-axiom
3. $(\varphi \wedge \psi) \leftrightarrow \langle \neg \pi \rangle \psi.$ 1, 2, SUBST.

A dual version of the derived negation rule:

$$\frac{\varphi \leftrightarrow \langle \pi \rangle \top}{(\varphi \vee \psi) \leftrightarrow [\neg \pi]\psi.}$$

A derived rule of implication.

$$\frac{\varphi \leftrightarrow [\pi_1]\psi \qquad \psi \leftrightarrow \langle \pi_2 \rangle \top}{(\varphi \wedge \chi) \leftrightarrow \langle \pi_1 \Rightarrow \pi_2 \rangle \chi.}$$

Its derivation:

4. $([\pi_1]\langle \pi_2 \rangle \top \wedge \chi) \leftrightarrow \langle \pi_1 \Rightarrow \pi_2 \rangle \chi$ $\Rightarrow$-axiom
1. $\varphi \leftrightarrow [\pi_1]\psi$ premiss
2. $\psi \leftrightarrow \langle \pi_2 \rangle \top$ premiss
3. $\varphi \leftrightarrow [\pi_1]\langle \pi_2 \rangle \top$ 1, 2, SUBST
4. $([\pi_1]\langle \pi_2 \rangle \top \wedge \chi) \leftrightarrow \langle \pi_1 \Rightarrow \pi_2 \rangle \chi$ $\Rightarrow$-axiom
5. $(\varphi \wedge \chi) \leftrightarrow \langle \pi_1 \Rightarrow \pi_2 \rangle \chi$ 4, 3, SUBST

A dual version of the derived rule of implication:

$$\frac{\varphi \leftrightarrow \langle \pi_1 \rangle \psi \qquad \psi \leftrightarrow [\pi_2]\bot}{(\varphi \vee \chi) \leftrightarrow [\pi_1 \Rightarrow \pi_2]\chi.}$$

A derived rule of η assignment:

$$\frac{\varphi \leftrightarrow \langle \pi \rangle \psi}{\exists x \varphi \leftrightarrow \langle \eta x : \pi \rangle \psi.}$$

Its derivation:

1. $\varphi \leftrightarrow \langle \pi \rangle \psi$ premiss
2. $\exists x \langle \pi \rangle \psi \leftrightarrow \langle \eta x : \pi \rangle \psi$ η-axiom
3. $\exists x \varphi \leftrightarrow \langle \eta x : \pi \rangle \psi$ 1, 2, SUBST

A dual version of the derived rule of η assignment:

$$\frac{\varphi \leftrightarrow [\boldsymbol{\pi}]\psi}{\forall x\varphi \leftrightarrow [\boldsymbol{\eta x} : \boldsymbol{\pi}]\psi.}$$

A derived rule of ι assignment:

$$\frac{\varphi \leftrightarrow \langle \boldsymbol{\pi} \rangle \top \qquad \psi \leftrightarrow \langle \boldsymbol{\pi} \rangle \chi}{(\exists! x\varphi \wedge \exists x\psi) \leftrightarrow \langle \boldsymbol{\iota x} : \boldsymbol{\pi} \rangle \chi.}$$

Its derivation:

1. $\varphi \leftrightarrow \langle \boldsymbol{\pi} \rangle \top$ premiss
2. $\psi \leftrightarrow \langle \boldsymbol{\pi} \rangle \chi$ premiss
3. $(\exists! x\langle \boldsymbol{\pi} \rangle \top \wedge \exists x\langle \boldsymbol{\pi} \rangle \chi) \leftrightarrow \langle \boldsymbol{\iota x} : \boldsymbol{\pi} \rangle \chi$ ι-axiom
4. $(\exists! x\varphi \wedge \exists x\langle \boldsymbol{\pi} \rangle \chi) \leftrightarrow \langle \boldsymbol{\iota x} : \boldsymbol{\pi} \rangle \chi$ 1, 3, SUBST
5. $(\exists! x\varphi \wedge \exists x\psi) \leftrightarrow \langle \boldsymbol{\iota x} : \boldsymbol{\pi} \rangle \chi$ 2, 4, SUBST

A dual version of the derived rule of ι assignment:

$$\frac{\varphi \leftrightarrow \langle \boldsymbol{\pi} \rangle \top \qquad \psi \leftrightarrow [\boldsymbol{\pi}]\chi}{(\exists! x\varphi \rightarrow \forall x\psi) \leftrightarrow [\boldsymbol{\iota x} : \boldsymbol{\pi}]\chi.}$$

3.7 Deriving Static Meanings of DPL Programs

We end this chapter with an example of a derivation of the static meaning of a DPL program in the calculus. We find the static meaning of a DPL program by computing a predicate logical formula which expresses the weakest preconditions under which the program succeeds. These weakest preconditions for success give the static truth conditions. Equivalently, we can compute static falsity conditions as weakest preconditions for failure of a program. Example (3.1) from Section 3.1 can be translated into DPL as follows.

$$(\eta x : Wx; \; Mx) \Rightarrow (\iota y : Hyx; \; Lyx). \tag{3.2}$$

We will derive the static truth conditions in two stages. First we use one of the derived rules for dynamic implication, which tells us that in order to find the formula φ we are looking for we need a formula ψ for which the following holds:

$$\frac{\vdots \qquad\qquad\qquad \vdots}{\varphi \leftrightarrow [\boldsymbol{\eta x} : \boldsymbol{Wx; Mx}]\psi \qquad \psi \leftrightarrow \langle \boldsymbol{\iota y} : \boldsymbol{Hyx; Lyx} \rangle \top}{\varphi \leftrightarrow \langle (\boldsymbol{\eta x} : \boldsymbol{Wx; Mx}) \Rightarrow (\boldsymbol{\iota y} : \boldsymbol{Hyx; Lyx}) \rangle \top.}$$

For the first subderivation we find:

$$\frac{(Wx \rightarrow (Mx \rightarrow \psi)) \leftrightarrow [\boldsymbol{Wx}](Mx \rightarrow \psi)}{\forall x(Wx \rightarrow (Mx \rightarrow \psi)) \leftrightarrow [\boldsymbol{\eta x} : \boldsymbol{Wx}](Mx \rightarrow \psi)} \qquad (Mx \rightarrow \psi) \leftrightarrow [\boldsymbol{Mx}]\psi}{\forall x(Wx \rightarrow (Mx \rightarrow \psi)) \leftrightarrow [\boldsymbol{\eta x} : \boldsymbol{Wx}; \boldsymbol{Mx}]\psi.}$$

The second subderivation computes ψ:

$$\frac{Hyx \leftrightarrow \langle \boldsymbol{Hyx} \rangle \top \qquad (Hyx \wedge Lyx)\langle \boldsymbol{Hyx} \rangle Lyx}{\frac{(\exists! yHyx \wedge \exists y(Hyx \wedge Lyx)) \leftrightarrow \langle \iota y : \boldsymbol{Hyx} \rangle Lyx \quad Lyx \langle \boldsymbol{Lyx} \rangle \top}{(\exists! yHyx \wedge \exists y(Hyx \wedge Lyx)) \leftrightarrow \langle \iota y : \boldsymbol{Hyx}; \boldsymbol{Lyx} \rangle \top.}}$$

Combining these two results we get that formula (3.7) expresses the static truth conditions of the example sentence.

$$\forall x(Wx \rightarrow (Mx \rightarrow (\exists! yHyx \wedge \exists y(Hyx \wedge Lyx)))). \tag{3.3}$$

For good measure we will also derive the weakest precondition φ for which the program fails. According to the other derived rule for dynamic implication this is the φ such that we can find a ψ for which the following holds:

$$\frac{\vdots \qquad\qquad \vdots}{\frac{\varphi \leftrightarrow \langle \boldsymbol{\eta x} : \boldsymbol{Wx}; \boldsymbol{Mx} \rangle \psi \qquad \psi[\iota y : \boldsymbol{Hyx}; \boldsymbol{Lyx}] \bot}{\varphi \leftrightarrow [(\boldsymbol{\eta x} : \boldsymbol{Wx}; \boldsymbol{Mx}) \Rightarrow (\iota y : \boldsymbol{Hyx}; \boldsymbol{Lyx})] \bot.}}$$

For the first subderivation we find:

$$\frac{(Wx \wedge Mx \wedge \psi) \leftrightarrow \langle \boldsymbol{Wx} \rangle (Mx \wedge \psi)}{\frac{\exists x(Wx \wedge Mx \wedge \psi) \leftrightarrow \langle \boldsymbol{\eta x} : \boldsymbol{Wx} \rangle (Mx \wedge \psi) \qquad (Mx \wedge \psi) \leftrightarrow \langle \boldsymbol{Mx} \rangle \psi}{\exists x(Wx \wedge Mx \wedge \psi) \leftrightarrow \langle \boldsymbol{\eta x} : \boldsymbol{Wx}; \boldsymbol{Mx} \rangle \psi.}}$$

The second subderivation computes ψ:

$$\frac{Hyx \leftrightarrow \langle \boldsymbol{Hyx} \rangle \top \qquad (Hyx \rightarrow \neg Lyx) \leftrightarrow [\boldsymbol{Hyx}]\neg Lyx}{\frac{(\exists! yHyx \rightarrow \forall y(Hyx \rightarrow \neg Lyx)) \leftrightarrow [\iota y : \boldsymbol{Hyx}]\neg Lyx \qquad \neg Lyx \leftrightarrow [\boldsymbol{Lyx}]\bot}{(\exists! yHyx \rightarrow \forall y(Hyx \rightarrow \neg Lyx)) \leftrightarrow [\iota y : \boldsymbol{Hyx}; \boldsymbol{Lyx}]\bot.}}$$

Combining these two results we get that formula (3.7) gives the static falsity conditions of the example sentence.

$$\exists x(Wx \land Mx \land (\exists! yHyx \to \forall y(Hyx \to \neg Lyx))). \tag{3.4}$$

Formula (3.7) expresses the negation of (3.7). This is what one would expect under the present regime. The uniqueness presupposition of the definite is swallowed up in the truth conditions. In [2, 3] a means is proposed for separating out the uniqueness presupposition by allowing the possibility of error abortion of a program in case a uniqueness presupposition is violated. This gets us a system where we can calculate static truth conditions, static falsity conditions and static error abortion conditions.

3.8 Conclusion

We have shown how dynamic predicate logic with definite assignment can be axiomatized in Pratt style quantified dynamic logic, and how the present calculus extends the Hoare style calculus for dynamic predicate logic from Van Eijck and De Vries [4]. Also, quantified dynamic logic turns out to be an excellent medium to formulate and discuss proposals for dynamic entailment relations.

Acknowledgements

This chapter has benefited from comments by Tim Fernando, Wilfried Meyer Viol, by the participants of the NFI workshop on 'Logic and the Flow of Information' (December 1991) and by the members of the Utrecht working group on dynamic and partial semantics. Special thanks go to Johan van Benthem for some particularly helpful suggestions.

References

[1]   K.R. Apt. Ten years of Hoare's logic: A survey—part i. *ACM Transactions on Programming Languages and Systems*, 3(4):431–483, 1981.

[2]   J. van Eijck. The dynamics of description. *Journal of Semantics*, 10:239–267, 1993.

[3]   J. van Eijck. Presupposition failure — a comedy of errors. Manuscript, CWI, Amsterdam, 1992.

[4]   J. van Eijck and F.J. de Vries. Dynamic interpretation and Hoare deduction. *Journal of Logic, Language, and Information*, 1:1–44, 1992.

[5]   H.B. Enderton. *A Mathematical Introduction to Logic*. Academic Press, 1972.

[6]   T. Fernando. Transition systems and dynamic semantics. Manuscript, CWI, Amsterdam, 1992.

[7]   R. Goldblatt. *Logics of Time and Computation, Second Edition, Revised and Expanded*, volume 7 of *CSLI Lecture Notes*. CSLI, Stanford, 1992 (first edition 1987). Distributed by University of Chicago Press.

[8]   J. Groenendijk and M. Stokhof. Dynamic predicate logic. *Linguistics and Philosophy*, 14:39–100, 1991.

[9]   C.A.R. Hoare. An axiomatic basis for computer programming. *Communications of the ACM*, 12(10):567–580, 583, 1969.

[10]   D. Kozen and J. Tiuryn. Logics of programs. In J. van Leeuwen, editor, *Handbook of Theoretical Computer Science, Volume B*, pages 789–840. Elsevier, 1990.

[11]   V. Pratt. Semantical considerations on Floyd–Hoare logic. *Proceedings 17th IEEE Symposium on Foundations of Computer Science*, pages 109–121, 1976.

[12]   B. Russell. On denoting. *Mind*, 14:479–493, 1905.

[13]   Y. Venema. *Many-dimensional modal logic*. PhD thesis, University of Amsterdam, 1992.

4   How Logic Emerges from the Dynamics of Information

Peter Gärdenfors

4.1   Introduction

It is often claimed that the *symbolic* approach to information processing is incompatible with connectionism and other associationist modes of representing information. I propose to throw new light on this debate by presenting two examples of how logic can be seen as emerging from an underlying information dynamics. The first example shows how intuitionistic logic results very naturally from an abstract analysis of the dynamics of information. The second example establishes that the activities of a large class of neural networks may be interpreted, on the symbolic level, as nonmonotonic inferences. On the basis of these examples I argue that symbolic and non-symbolic approaches to information can be described in terms of different perspectives on the same phenomenon. Thus, I find that Fodor and Pylyshyn's claim that connectionist systems cannot be systematic and compositional is based on a misleading interpretation of representations in such systems.

4.2   Two Paradigms of Cognitive Science

There are currently two dominating paradigms concerning how cognitive processes can be identified. The first is the symbolic paradigm according to which the atoms of mental representations are symbols which combine to form meaningful expressions. Information processing involves above all computations of logical consequences. In brief, the mind is seen as a Turing machine that operates on sentences from a mental language by symbol manipulation. The second paradigm claims that cognition is characterized by *associations*. This idea goes back to the empiricist philosophers, but it has recently seen a revival in the emergence of *connectionism*. It has been often been claimed that these two paradigms are fundamentally at odds with one another, most notably by Fodor and Pylyshyn [15]. I shall argue that they are not.

4.2.1   The Symbolic Paradigm

The central tenet of the symbolic paradigm is that mental representation and information processing is essentially *symbol manipulation*. The symbols can be concatenated to form expressions in a *language of thought* – sometimes called Mentalese. A *mental state* is identified with a set of attitudes towards such expressions.

The content of a sentence in Mentalese is a proposition or a thought of a person. The different propositional attitudes in the mental states of a person are connected via their *logical* or *inferential relations*. Pylyshyn writes [29, p. 194]:

"If a person believes (wants, fears) P, then that person's behavior depends on the form the expression of P takes rather than the state of affairs P refers to ..."

In applications within AI, first order logic has been the dominating inferential system, but in other areas more general forms of inference, like those provided by statistical inference, inductive logic or decision theory, have been utilized.

Processing the information contained in a mental state consists in computing the consequences of the propositional attitudes, using some set of *inference rules*. The following quotation from Fodor [13, p. 230] is a typical formulation of the symbolic paradigm:

"Insofar as we think of mental processes as computational (hence as formal operations defined on representations), it will be natural to take the mind to be, inter alia, a kind of computer. That is, we will think of the mind as carrying out whatever symbol manipulations are constitutive of the hypothesized computational processes. To a first approximation, we may thus construe mental operations as pretty directly analogous to those of a Turing machine."

The material basis for these processes is irrelevant to the description of their results - the same mental state with all its propositional attitudes can be realized in a brain as well as in a computer. Thus, the symbolic paradigm clearly presupposes a *functionalist* philosophy of mind. The inference rules of logic and the electronic devices which conform to these rules are seen to be analogous to the workings of the brain. In brief, the mind is thought to be a computing device, which generates symbolic sentences as inputs from sensory channels, performs logical operations on these sentences, and then transforms them into linguistic or non-linguistic output behaviors.

A further claim of the symbolic paradigm is that mental representations *can not be reduced* to neurobiological categories. The reason is that the functional role of the symbolic representations and the inference rules can be given many different realizations, neurophysiological or others. The causal relations governing such a material realization of a mental state will be different for different realizations, even if they represent the same logical relations. Thus, according to functionalism, the logical relations that characterize mental representations and the information processing cannot be reduced to any underlying neurological or electronic causes. (Cf. Churchland [7, Ch. 9.5])

The outline of the symbolic paradigm that has been presented here will not be explicitly found in the works of any particular author. However, a defense of the general reasoning can be found, for example, in the writings of Jerry Fodor [13, Introduction, and Chs. 7 and 9] and Zenon Pylyshyn [29], and in their joint article [15]. It is also clear that the symbolic paradigm forms an implicit methodology for most of the research in AI.

4.2.2   The Associationist Paradigm

For Hume, thinking consists basically in the forming of *associations* between "perceptions of the mind." This idea has since then been developed by the British empiricists, the American pragmatists (William James), and, in particular, by the behaviorists. Their stimulus-response pairs are prime examples of the notion of an association. Dellarosa [9, p. 29] summarizes the central tenet as follows:

"Events that co-occur in space or time become connected in the mind. Events that share meaning or physical similarity become associated in the mind. Activation of one unit activates others to which it is linked, the degree of activation depending on the strength of association. This approach held great intuitive appeal for investigators of the mind because it seems to capture the flavor of cognitive behaviors: When thinking, reasoning, or musing, one thought reminds us of others."

During the last decades, associationism has been revived with the aid of a new model of cognition: *connectionism*. Connectionist systems, also called neural networks, consist of large numbers of simple but highly interconnected units ("neurons"). Each unit receives activity, both excitatory and inhibitory, as input; and transmits activity to other units according to some function (normally non-linear) of the inputs. The behavior of the network as a whole is determined by the initial state of activation and the connections between the units. The inputs to the network also change the 'weights' of the connections between units according to some learning rule. Typically, the change of connections is much slower than changes in activity values. The units have no memory of themselves, but earlier inputs may be represented indirectly via the changes in weights they have caused.[1] In the literature one finds several different kinds of connectionist models (Rumelhart and McClelland [31], Beale and Jackson [3], Zornetzer, Davis and Lau [41]) that can be classified according to their architecture or their learning rules.

Connectionist systems have become popular among psychologists and cognitive scientists since they seem to be excellent tools for building models of associationist theories. And networks have been developed for many different kinds of tasks, including vision, language processing, concept formation, inference, and motor control (Beale and Jackson [3], Zornetzer, Davis and Lau [41]). Among the applications, one finds several that traditionally were thought to be typical symbol processing tasks. In favor of the neural networks, it is claimed by the connectionists that these models do not suffer from the brittleness of the symbolic models and that they are much less sensitive to noise in the input (Rumelhart and McClelland [31]).

---

[1] For a more formal treatment of neural networks, see Section 4.4.1.

4.2.3   The Unification Program: Different Perspectives on Information

It is generally claimed that the symbolic and the associationist/connectionist paradigms are *incompatible*. Some of the most explicit arguments for this position have been put forward by Smolensky [35, p. 7] and Fodor and Pylyshyn [15]. Smolensky argues that, on the one hand, symbolic programs requires linguistically formalized precise rules that are sequentially interpreted (hypothesis 4a in his paper); and, on the other hand, connectionist systems cannot be given a complete and formal description on the symbolic level (hypothesis 8).

He also rebuts the argument that, in principle, one type of system can be *simulated* by a system of the other kind. Firstly, he argues, connectionist models cannot be "merely implementations, for a certain kind of parallel hardware, of symbolic programs that provide exact and complete accounts of behavior at the conceptual level" (Hypothesis 10, p. 7) since this conflicts with the connectionist assumption that neural networks cannot be completely described on the symbolic ("conceptual") level (Hypothesis 8c, pp. 6–7). Secondly, even if a symbolic system is often used to implement a connectionist system, "the symbols in such programs represent the activation values of units and the strength of connections" (p. 7), and they do not have the conceptual semantics required by the symbolic paradigm. Thus the translated programs are not symbolic programs of the right kind.

Fodor and Pylyshyn's [15] main argument for the incompatibility of the symbolic and the associationist/connectionist paradigms is that connectionist models, in contrast to the symbolic, lack *systematicity* and *compositionality*.[2] By saying that the capabilities of a system are systematic they mean that "the ability to produce/understand some sentences [symbolic expressions] is *intrinsically* connected to the ability to produce/understand certain others." [ibid., p. 37] To give a simple example, if you can express or represent "Abel hits Cain" and "Beatrice kisses David," you can also express or represent, e.g., "Cain hits David" and "Abel kisses Beatrice." Compositionality is a well-known principle which requires that symbolic expressions can be composed into new meaningful expressions. This principle is required to "explain how a finitely representable language can contain infinitely many nonsynonymous expressions." [ibid., p. 43] Fodor and Pylyshyn's arguments for why connectionist systems cannot be systematic and compositional will be presented (and criticized) in Section 4.5.2.

It thus seems that there is an impenetrable wall between the symbolic and the connectionist paradigms. One of my aims in this article is to show that they can be unified. I will argue that the alleged conflict between these paradigms can be resolved by adopting two different *perspectives* on how information is processed in various systems.

---

[2]They also claim that they lack *productivity* (Section 3.1 in their paper) and *inferential coherence* (Section 3.4), but these arguments seem to carry less weight for them.

One perspective on an information processing system is to look at its *dynamics*, i.e., how one state of the system is transformed to another, given a particular input to the system. This perspective is the normal one to use when describing a connectionist system. The other perspective is to forget about the details of the transition from the input to the output and only consider what is represented by the input and its relation to what is represented by the output. As will be shown below, this relation can often be interpreted as a symbolic *inference*, completely in accordance with the requirements of the symbolic paradigm. Thus, in a sense to be made more precise later, by changing from one perspective to the other, symbolic inferences can be seen as *emerging* from dynamic 'associations'. The pivotal point is that there is no need to distinguish between two kinds of systems – the two perspectives can be adopted on a single information processing system.

I will start out, in Section 4.3, by describing the dynamics of an information processing system in a very abstract way. Here the details of the process are of no importance. What counts is merely the relation between the input and the output. Using this simple structure, I shall introduce a definition of what a *proposition* is, which does not presume any form of symbolic structure. Nevertheless, if one looks upon these propositions from another perspective, it turns out that there is a *logic* to them.

In Section 4.4, I will become more concrete and actually use connectionist systems as models of the dynamic processes. Again, by adopting a different perspective on what the system is doing, I shall show that it can be seen as performing logical inferences. It turns out that nonmonotonic inferences can be modelled in a natural fashion by such systems.

In Section 4.5, I shall return to the alleged clash between the symbolic and the connectionist paradigms. In the light of the example from Section 4.4, I shall argue that Fodor and Pylyshyn's criticism of connectionist systems is misplaced. Furthermore, I shall use the examples from Sections 4.3 and 4.4 to support the claim that there is no fundamental conflict between the two views.

4.3   The Dynamics of Information as a Basis for Logic

The proper objects of logic are not sentences but the *content* of sentences. Thus, in order to understand what logic is about, one needs a theory of propositions. In traditional philosophical logic, a proposition is often defined in terms of possible worlds, so that a proposition is identified with the set of worlds in which it is true.

With this definition, it is easy to see how the *logic* of propositions can be generated. By using standard set-theoretical operations, we can form composite propositions: The conjunction of two propositions is represented by the intersection of the sets of possible worlds representing the propositions; the disjunction is represented by their union; the negation is represented by the complement with respect to the set of all possible worlds;

etc. As is easily seen, this way of constructing the standard logical connectives results in *classical* truth-functional logic. The underlying reason is simply that the 'logic' of the set-theoretical operations is classical. In this sense we see how already the *ontology* used when defining propositions determines their logic.

4.3.1 An Alternative Definition of Propositions

I shall now present another way of defining a proposition, based on the *dynamics of information states*, and show how this definition leads to a different perspective on logic. The construction presented here is adapted from Gärdenfors [17, 18].

The ontologically fundamental entities in the reconstruction of a propositional logic will be *information states*.[3] In this section, no assumptions whatsoever will be made about the structure of the information states. However, the interpretation is that they represent states in *equilibria* in the sense that the underlying dynamic processes are assumed to have operated until a stable state is reached.[4]

What can change an information state in equilibrium is that it receives some form of informational input[5] that upsets the equilibrium and starts the dynamic processes again. Here, I will avoid all problems connected with a more precise description of *what* an informational input is. I will simply identify an input with the change it induces in an information state.

Formally, this idea can be expressed by defining an informational input as a *function* from information states to information states. When a function representing a certain input is applied to a given information state K, the value of the function is the state which would be the result of accommodating the input to K. This way of defining informational input via changes of belief is analogous to defining events via changes of physical states.

If two inputs always produce the same new information state, i.e., if the inputs are identical as functions, there is no epistemological reason to distinguish them. Apart from information states, the only entities that will be assumed as primitives in this section are functions of this kind which take information states as arguments and values.

The most important type of input is when new evidence is accepted as certain or 'known' in the resulting information state. Below, I will concentrate on this type of input.[6] Input corresponding to accepting evidence as certain represents the simplest kind of expanding an information state and is one way of modelling learning. The information contained in such an input will be called a *proposition*. Following the general identification of inputs presented above, *propositions are defined as functions from information states to information states*. This definition will be the point of departure for

---

[3]Information states were called states of belief in Gärdenfors [17, 19].
[4]Cf. the 'resonant states' described in Section 4.4.1.
[5]Informational inputs were called epistemic inputs in Gärdenfors [17, 19].
[6]Other forms of informational inputs are discussed in Gärdenfors [19].

the reconstruction of propositional logic from the dynamics of information. Veltman [39, p. 1] gives an informal account of this definition in the following way: "You know the meaning of a sentence if you know the change it brings about in the information state of anyone who accepts the news conveyed by it."

As a first application of the definition, a central concept for information can now be introduced: A proposition A is said to be *accepted as known in the information state K* if and only if $A(K) = K$.

It is important to keep in mind that not all functions that can be defined on information states are propositions. Propositions correspond to a certain type of informational input, to wit, when new evidence is accepted as certain. In order to characterize the class of propositions, I will next formulate some postulates for propositions. Before this is done, we cannot speak of the *logic* of propositions.

4.3.2 Basic Postulates for Propositions

First we need a definition of the basic dynamic structure. A *dynamic model*[7] is a pair $\langle \mathbf{K}, \mathbf{P} \rangle$, where $\mathbf{K}$ is a set and $\mathbf{P}$ is a class of functions from $\mathbf{K}$ to $\mathbf{K}$. Members of $\mathbf{K}$ will be called information states and they will be denoted $K, K', \ldots$ The elements in $\mathbf{P}$ represent the propositions. $A, B, C, \ldots$ will be used as variables over $\mathbf{P}$. It should be noted once again that nothing is assumed about the structure of the elements in $\mathbf{K}$.

It will be assumed that the informational inputs corresponding to propositions can be iterated and that the composition of two such inputs is also a proposition. The composition of two propositions A and B will be denoted $A \wedge B$ (remember that $A \wedge B$ is not an element of some formal language, but a function from information states to information states). Formally, this requirement is expressed in the following postulate:

P 1   *For every A and B in* **P**, *there is a function* $A \wedge B$ *which is also in* **P** *such that, for every K in* **K**, $A \wedge B(K) = A(B(K))$.

It will also be postulated that the composition operation is commutative and idempotent:[8]

P 2   *For every A and B in* **P** *and for every K in* **K**, $A \wedge B(K) = B \wedge A(K)$.

P 3   *For every A in* **P** *and for every K in* **K**, $A \wedge A(K) = A(K)$.

We can now introduce the fundamental relation of *logical consequence* between propositions: A proposition B is a consequence of a proposition A in a dynamic model $\langle \mathbf{K}, \mathbf{P} \rangle$, if and only if $B(A(K)) = A(K)$, for all K in $\mathbf{K}$.

The identity function, here denoted by $\top$, will be assumed to be a proposition:

---

[7]Dynamic models were called belief models in Gärdenfors [19].
[8]For motivations of these and the following postulates, see Gärdenfors [17].

P 4   *The function* ⊤, *defined by* ⊤$(K) = K$, *for all* K *in* **K**, *is in* **P**.

A proposition A is a *tautology* in the dynamic model $\langle \mathbf{K}, \mathbf{P} \rangle$, if and only if $A(K) = ⊤(K)$, for all K in **K**.

The next postulate will be a formal characterisation of the information obtained when one learns that one thing *implies* another (or is *equivalent* to another).

P 5   *For every A and B in* **P**, *there is a function C in* **P** *such that*

a.   *for all K in* **K**, $A(C(K)) = B(C(K))$;

b.   *for any function D in* **P** *such that* $A(D(K)) = B(D(K))$ *for all K in* **K**, *there is a function E in* **P** *such that* $D(K) = E(C(K))$, *for all K in* **K**.

A function C which satisfies (a) and (b) will be called an *equalizer* of A and B. With the aid of (P2) it can be shown that there is only one equalizer of A and B in **P**. We can thus give the proposition postulated in (P5) a well defined name: the equalizer of A and B will be denoted $A \leftrightarrow B$. From this we define the proposition $A \rightarrow B$ which corresponds to the information that A implies B, by the equation $(A \rightarrow B)(K) = (A \leftrightarrow (A \wedge B))(K)$. for all K in **K**.

The negation of a proposition will be defined by first assuming the existence of a 'falsity' proposition:

P 6   In every dynamic model $\langle \mathbf{K}, \mathbf{P} \rangle$ there exists a constant function $\bot$ in **P**, i.e., there is some $K_\bot$ in **K** such that $\bot(K) = K_\bot$ for all K in **K**.

$K_\bot$ will be called the *absurd* information state. As is standard in propositional logic, the negation $\neg A$ of a proposition A is defined as the proposition $A \rightarrow \bot$.

The postulate for disjunction will be in the same style as the postulate concerning equalizers:

P 7   *For every A and B in* **P**, *there is a function C in* **P** *such that*

a.   *for all K in* **K**, $A(C(K)) = A(K)$ *and* $B(C(K)) = B(K)$;

b.   *for any function D in* **P** *that satisfies (a), there is a function E in* **P** *such that* $E(D(K)) = C(K)$ *for all K in* **K**.

A function C that satisfies (a) and (b) will be called a *disjunction* of A and B.

4.3.3   Completeness Results

I have now introduced postulates for operations on propositions that correspond to each of the standard propositional connectives. I will next present some technical results which answer the question of which 'logic' is determined by these postulates.

The crucial point in my construction is that *expressions* like $(A \leftrightarrow (A \wedge B))(K)$ *can be viewed from two perspectives.* Officially, the expressions like $A \leftrightarrow (A \wedge B)$ are not sentences in a language but *functions* defined on information states. However, given that the postulates (P1)–(P7) are satisfied for the class of functions in **P** there is, of course, an obvious one-one correspondence between the propositions in a dynamic model and the sentences in a standard 'propositional' language ($\top$ and $\perp$ are sentential constants in this language). In other words, when (P1)–(P7) are satisfied, a syntactic structure 'emerges' from the class of functions. This entails that we can consider the propositions that are tautologies in a given dynamic model as a class of sentences and then ask how the formulas which are included in all such classes can be axiomatized. If we really want to have an explicit symbolic structure, we can view expressions of the form $A \leftrightarrow (A \wedge B)$ as *names* of the functions. The point is that the referents of the names are uniquely determined by the names themselves.[9] In other words, the *semantics* of the language is trivial: it is the identity mapping. However, in this mapping the same object is given a double interpretation: on the one hand it is a symbolic expression in a formal language; on the other, it is a function in a dynamic model.

It can be shown Gärdenfors [18] that the logic generated by the postulates (P1)–(P7) is *exactly* intuitionistic logic.[10] In order to derive classical propositional logic, we need one more postulate for the class of propositions in a dynamic model:

P 8 *For every A and B in* **P**, $A \leftrightarrow B = \neg A \leftrightarrow \neg B.$

In this section, I have shown how propositional logic can be constructed from informational dynamics. The key idea for the construction is the definition of a proposition as a function representing changes of belief. The ontological basis of the construction is very meagre. The only entities that have been assumed to exist are information states and functions defined on information states. In order to emphasize this further, let me mention some things that have *not* been assumed: Firstly, it is not necessary that there be an independent object language that expresses the propositions to be studied. In contrast, the structure of an appropriate language emerges from the class of functions when the postulates are satisfied. Secondly, no set theory has been used; all constructions have been expressed solely in terms of functions. Thirdly, it can be noted that the construction does not use the concept of truth or possible worlds in any way.

---

[9]It is interesting to note that the key idea behind a Henkin completeness proof is based on the same kind of identification: The objects in the Henkin models are determined from equivalence classes of formulas.

[10]This follows essentially from the fact that any pseudo-Boolean (Heyting) algebra for intuitionistic logic can be used to construct a dynamic model which is equivalent to the pseudo-Boolean algebra in the sense that an element is identical with the unit element in the pseudo-Boolean algebra if and only if the corresponding proposition is a tautology in the dynamic model (this construction is presented in Gärdenfors [18]. Cf. van Benthem [4] for further connections between various kinds of algebras and dynamic models.

The dynamic approach to logic presented in this section has been generalized in a several ways by a number of logicians in Amsterdam. Veltman [39] extends it to an analysis of the function of 'might' and to default rules in general. Groenendijk and Stokhof [23] provide a dynamic interpretation of first-order predicate logic. Their dynamic predicate logic can be seen as a compositional, non-representational discourse semantics. Apart from giving a dynamic analysis of quantifiers, they show in particular how this approach can be used to handle anaphoric reference. They also compare it to Kamp's [25] discourse representation theory. In Groenendijk en Stokhof [22], they extend their approach to a typed language with λ-abstraction and use it to supply a semantic component for a Montague-style grammar. Van Eijck and de Vries [11] use the approach of Groenendijk and Stokhof and extend dynamic predicate logic with ι-assignments and with generalized quantifiers. Again, their semantics is applied to problems of anaphoric reference. Van Benthem [4] adopts a very general approach and discusses a number of ways of connecting a dynamic approach to logic with other more traditional logical and algebraic theories. A special case of this is the dynamic modal logic DML developed in de Rijke [30].

4.4   Neural Networks as Nonmonotonic Inference Machines

In physical systems one often finds descriptions of 'slow' and 'fast' aspects of dynamic processes. A well-known example from statistical mechanics is the 'slow' changes of temperature as a different perspective on a complex system of 'fast' moving gas molecules. Another example is catastrophe theory (Thom [38]) which is an entire mathematical discipline devoted to investigating the qualitative properties (in particular the 'catastrophes') of the 'slow' manifolds generated by a dynamical system.

The analogy I want to make in the context of the conflict between the symbolic and the associationist paradigms is that associationism deals with the 'fast' behavior of a dynamic system, while the symbolic structures may emerge as 'slow' features of such a system. In particular, inferential relations can be described from both perspectives. The upshot is that one and the same system, depending on the perspective adopted, can be seen as both an associationist mechanism and as an inferential rule-following process operating on symbolic structures.

In this section I shall elaborate on this double interpretation for the case when the system is a *neural network*.[11] Pictorially, the 'fast' behavior of a neural network are the 'associations' between the neurons in the network, i.e., the transmission of the activity levels within the network. In other words, what the network does is to locate minima in a 'cognitive dissonance function' (which, e.g., can be identified as maxima in Smolensky's [34] harmony functions). I want to argue that the corresponding 'slow' behavior of many networks can be described as the results of the network performing *inferences* in a

---

[11]This section is, to a large extent, borrowed from Balkenius and Gärdenfors [2].

precisely defined sense, and with a well-defined logical structure. It turns out the these inferences are, in a very natural way, *nonmonotonic*.

It should be emphasized that there is a different, even slower, process in a neural network, namely the *learning* that occurs from new instances being presented to the system and which causes the connections between the neurons to change. As is argued in Gärdenfors [20], this kind of change within a neural network corresponds to another kind of inference, to wit, *inductive* inferences. In this chapter, however, I will not consider learning processes in neural networks.

4.4.1 Schemata and Resonant States in Neural Networks

First of all we need a general description of neural networks. One can define a neural network N as a 4-tuple $\langle S, F, C, G \rangle$. Here S is the space of all possible *states* of the neural network. The dimensionality of S corresponds to the number of parameters used to describe a state of the system. Usually $S = [a, b]^n$, where $[a, b]$ is the working range of each neuron and n is the number of neurons in the system. We will assume that each neuron can take excitatory levels between 0 and 1. This means that a state in S can be described as a vector $x = \langle x_1, \ldots, x_n \rangle$ where $0 \leq x_i \leq 1$, for all $1 \leq i \leq n$. The network N is said to be binary if $x_i = 0$ or $x_i = 1$ for all i, that is if each neuron can only be in two excitatory levels.

C is the set of possible *configurations* of the network. A configuration $c \in C$ describes for each pair i and j of neurons the connection c_{ij} between i and j. The value of c_{ij} can be positive or negative. When it is positive the connection is excitatory and when it is negative it is *inhibitory*. A configuration c is said to be symmetric if $c_{ij} = c_{ji}$ for all i and j.

F is a set of *state transition functions* or *activation functions*. For a given configuration $c \in C$, a function $f_c \in F$ describes how the neuron activities spread through that network. G is a set of *learning functions* which describe how the configurations develop as a result of various inputs to the network.

By changing the behavior of the functions in the two sets F and G, it is possible to describe a large set of different neural mechanisms. In the rest of the section, I will assume that the state in C is fixed while studying the state transitions in S. This means that I will not consider the effects of learning within a neural network.

In Balkenius [1990] and Balkenius and Gärdenfors [2] it is argued that there is a very simple way of defining the notion of a *schema* within the theory of neural networks that can be seen as a generalization of the notion of a proposition. The definition proposed there is that a schema α corresponds to a vector $\langle \alpha_1, \ldots, \alpha_n \rangle$ in the state space S. That a schema α is currently *represented* (or *accepted*) in a neural network with an activity vector $x = \langle x_1, \ldots, x_n \rangle$ means that $x_i \leq \alpha_i$, for all $1 \leq i \leq n$. There is a natural way of defining a partial order of 'greater informational content' among schemata by putting

$\alpha \leq \beta$ iff $a_i \leq \beta_i$ for all $1 \leq i \leq n$. There is a minimal schema in this ordering, namely $\mathbf{0} = \langle 0, \dots, 0 \rangle$ and a maximal element $\mathbf{1} = \langle 1, \dots, 1 \rangle$.

In the light of this definition, let us consider some general desiderata for schemata. First, it is clear that depending on what the activity patterns in a neural network correspond to, schemata as defined here can be used for representing objects, situations, and actions.[12]

Second, if $\alpha \geq \beta$, then β can be considered to be a more *general* schema than α and α can thus be seen as an *instantiation* of the schema β. The part of α not in β, is a *variable* instantiation of the schema β. This implies that all schemata with more information than β can be considered to be an instantiation of β with different variable instantiations. Thus, schemata can have variables even though they do not have any *explicit* representation of variables.[13] Only the *value* of the variable is represented and not the variable as such. In general, it can be said that the view on schemata presented here replaces symbols by vectors representing various forms of *patterns*.

Third, it will soon be shown that schemata support *default assumptions* about the environment. The neural network is thus capable of filling in missing information.

The abstract definition of schemata presented here fits well with Smolensky's [34] analysis of schemata in terms of 'peaks' in a harmony function. And in Smolensky [36, p. 202] he says that his treatment of connectionism

"is committed to the hypothesis that mental representations are vectors partially specifying the state of a dynamical system (the activities of units in a connectionist network), and that mental processes are specified by the dynamical equations governing the evolution of that dynamical system."

Some interesting examples of schemata are found in Rumelhart, Smolensky, McClelland and Hinton [32] who address, among other things, the case of schemata for rooms. The network they investigate contains 40 neurons representing microfeatures of rooms like has- ceiling, contains-table, etc.[14] There are no neurons in the network representing kitchens and bedrooms, but various rooms can be represented implicitly as schemata of the network; the peaks of the harmony function correspond to prototypical rooms.

At this point, I want to emphasize that the definition of schemata given here is the simplest possible and is introduced just to show that even with elementary means it is possible to exhibit the compositional and systematic structure desired by the adherents of the symbolic paradigm. The definition applies to any neural network falling under

---

[12]For some examples of this, cf. Balkenius [1].

[13]Smolensky's [37] solution to the problem of variables is more complicated and to some extent ad hoc. On the other hand, he can handle asymmetric relations and some embedding features that cannot be given a simple analysis on the present approach.

[14]The network is presented on pp. 22–24 in Rumelhart, Smolensky, McClelland and Hinton [32].

the general description above. However, for networks that are designed for some special purpose, it is possible to introduce more sophisticated and fine-structured definitions of schemata that better capture what the network is intended to represent.

There are some elementary operations on schemata as defined above that will be of interest when I consider nonmonotonic inferences in a neural network. The first operator is the *conjunction* $\alpha \bullet \beta$ of two schemata $\alpha = \langle \alpha_1, \ldots, \alpha_n \rangle$ and $\beta = \langle \beta_1, \ldots, \beta_n \rangle$ which is defined as $\langle \gamma_1, \ldots, \gamma_n \rangle$, where $g_i = max(\alpha_i, \beta_i)$ for all i. If schemata are considered as corresponding to observations in an environment, one can interpret $a \bullet b$ as the *coincidence* of two schemata, i.e., the simultaneous observation of two schemata.

Secondly, the *complement* α^* of a schema $\alpha = \langle \alpha_1, \ldots \alpha_n \rangle$ is defined as $\langle 1 - \alpha_1, \ldots, 1 - \alpha_n \rangle$ (recall that 1 is assumed to be the maximum activation level of the neurons, and 0 the minimum). In general, the complementation operation does not behave like negation since, for example, if $\alpha = \langle 0.5, \ldots, 0.5 \rangle$, then $\alpha^* = \alpha$. However, if the neural network is assumed to be binary, that is if neurons only take activity values 1 or 0, then * will indeed behave as a classical negation on the class of binary-valued schemas.

Furthermore, the interpretation of the complement is different from the classical negation since the activities of the neurons only represent *positive* information about certain features of the environment. The complement α^* reflects a lack of positive information about α. It can be interpreted as a schema corresponding to the observation of everything but α. As a consequence of this distinction it is pointless to define implication from conjunction and complement. The intuitive reason is that it is impossible to observe an implication directly. A consequence is that the ordering *geq* only reflects greater positive informational content.

Finally, the *disjunction* $\alpha \oplus \beta$ of two schemata $\alpha = \langle \alpha_1, \ldots, \alpha_n \rangle$ and $\beta = \langle \beta_1, \ldots, \beta_n \rangle$ is defined as $\langle \gamma_1, \ldots, \gamma_n \rangle$, where $\gamma_i = min(\alpha_i, \beta_i)$ for all i. The term 'disjunction' is appropriate for this operation only if we consider schemata as representing propositional information. Another interpretation that is more congenial to the standard way of looking at neural networks is to see α and β as two instances of a *variable*. $\alpha \oplus \beta$ can then be interpreted as the *generalization* from these two instances to an underlying variable.

It is trivial to verify that the De Morgan laws $\alpha \oplus \beta = (\alpha^* \bullet \beta^*)^*$ and $\alpha \bullet \beta = (\alpha^* \oplus \beta^*)^*$ hold for these operations. The set of all schemata forms a distributive lattice with zero and unit, as is easily shown. It is a boolean algebra, whenever the underlying neural network is binary. In this way we have already identified something that can be viewed as a *syntactic* (and *compositional*) structure on the set of *vectors* representing schemata.

How does the structure on states of networks, generated by the operators $\bullet$, $\oplus$, and * relate to the postulates (P1)–(P7) in Section 4.3? For each schema α, it is trivial to define a function on activity states $x = \langle x_1, \ldots, x_n \rangle$ of a network, corresponding to giving α as an input, by putting $\alpha(x) = \alpha \bullet x$, for all x in S. Then if we put $\wedge = \bullet$, $\vee = \oplus$, $\top = 0$ and $\perp = 1$ it is easy to verify that postulates (P1)–(P4) and (P6)–(P7) hold. On the other hand, there does not seem to be any operator on vectors that is an equalizer. The candidate from classical logic, i.e., $(\alpha \bullet \beta) \oplus (\alpha^* \bullet \beta^*)$, does not satisfy the requirements of (P5) in general. However, in the special case when the network is binary, this construction works and all of the postulates (P1) - (P8) are satisfied.

4.4.2   Nonmonotonic Inferences in a Neural Network

A desirable property of a network that can be interpreted as performing *inferences* of some kind is that it, when given a certain input, stabilizes in a state containing the results of the inference. In the theory of neural networks such states are called *resonant states*. In order to give a precise definition of this notion, consider a neural network $N = \langle S, F, C, G \rangle$. Let us assume that the configuration c is fixed so that we only have to consider one state transition function $f = f_c$. Let $f^0(x) = f(x)$ and $f^{n+1}(x) = f \circ f^n(x)$. Then a state y in S is called *resonant* if it has the following properties:

i. $f(y) = y$ (equilibrium)

ii.   If for any $x \in S$ and each $\epsilon > 0$ there exists a $\delta > 0$ such that $\mid x - y \mid < \delta$, then $\mid f^n(x) - y \mid < \epsilon$ when $n \geq 0$ (stability)

iii.   There exists a δ such that if $\mid x - y \mid < \delta$, then $\lim_{n \to \infty} f^n(x) = y$ (asymptotic stability).

Here $\mid . \mid$ denotes the standard euclidean metric on the state space S. A neural system N is called *resonant* if for each x in S there exists a $n > 0$, that depends only on x, such that $f^n(x)$ is a resonant state.

If $\lim_{n \to \infty} f^n(x)$ exists, it is denoted by $[x]$ and $[.]$ is called the *resonance function* for the configuration c. It follows from the definitions above that all resonant systems have a resonance function. For a resonant system we can then define *resonance equivalence* as $x \sim y$ iff $[x] = [y]$. It follows that $\sim$ is an equivalence relation on S that partitions S into a set of equivalence classes.

It can be shown (Cohen and Grossberg [8], Grossberg [24]) that a large class of neural networks have resonance functions. A common feature of these types of neural networks is that they are based on *symmetrical* configuration functions c, that is, the connections between two neurons are equal in both directions.

The function $[.]$ can be interpreted as filling in *default* assumptions about the environment, so that the schema represented by $[\alpha]$ contains information about what the network expects to hold when given α as input. Even if α only gives a partial description of, for example, an object, the neural network is capable of supplying the missing information in attaining the resonant state $[\alpha]$. The expectations are determined by the configuration function c, and thus expectations are 'equilibirum' features of a network in contrast to the transient input α.

I now turn to the problem of providing a different perspective of the activities of a neural network which will show it to perform *nonmonotonic inferences*. A first idea for describing the nonmonotonic inferences performed by a neural network N is to say that the resonant state $[\alpha]$ represents the expectations of the network given the input information represented by α. The expectations can also be described as the set of nonmonotonic conclusions to be drawn from α. However, the schema α is not always included in $[\alpha]$, that is, $[\alpha] \geq \alpha$ does not hold in general. Sometimes a neural network *rejects* parts of the input information – in pictorial terms it does not always believe what it sees.

So if we want α to be included in the resulting resonant state, one has to modify the definition. The most natural solution is to 'clamp' α in the network, that is to add the *constraint* that the activity levels of all neurons is above α_i, for all i. Formally, we obtain this by first defining a function f_α via the equation $f_\alpha(x) = f(x) \bullet \alpha$ for all $x \in S$. We can then, for any resonant system, introduce the function $[.]^\alpha$ (for a fixed configuration $c \in C$) as follows:

$$[x]^\alpha = \lim_{n \to \infty} f_\alpha n(x)$$

This function will result in resonant states for the same neural networks as for the function $[.]$.

The key idea of this section is then to define a nonmonotonic inference relation $\vdash\!\sim$ between schemata in the following way:

$$\alpha\vdash\!\sim\beta \text{ iff } [\alpha]^\alpha \geq \beta$$

This definition fits very well with the interpretation that nonmonotonic inferences are based on the dynamics of information as developed in Section 4.3. Note that α and β in the definition of $\vdash\!\sim$ have double interpretations: From the *connectionist* perspective, they are schemata which are defined in terms of activity vectors in a neural network. From the other perspective, the *symbolic*, they are viewed as formal expressions with a grammatical structure. Thus, in the terminology of Smolensky [35], we make the transition from the subsymbolic level to the symbolic simply by giving a different *interpretation* of the structure of a neural network. Unlike Fodor and Pylyshyn [15], we need not assume two different systems handling the two different levels. In contrast, the symbolic level *emerges* from the subsymbolic in one and the same system.

Before turning to an investigation of the general properties of $\vdash\!\sim$ generated by the definition, I will illustrate it by showing how it operates for a simple neural network.

EXAMPLE The network depicted below consists of four neurons with activities $x_1, \ldots, x_4$. Neurons that interact are connected by lines. Arrows at the ends of the lines indicate that the neurons excite each other; dots indicate that they inhibit each other. If we consider only schemata corresponding to binary activity vectors, it is possible to identify schemata with sets of active neurons. Let three schemata α, β, and γ correspond to the following activity vectors: $\alpha = \langle 1\,1\,0\,0 \rangle$, $\beta = \langle 0\,0\,0\,1 \rangle$, $\gamma = \langle 0\,1\,1\,0 \rangle$. Assume that x_4 inhibits x_3 more than x_2 excites x_3. If a is given as input, the network will activate x_3 and thus γ. It follows that $[\alpha]^\alpha \geq \gamma$ and hence $\alpha\!\mid\!\sim\!\gamma$. Extending the input to $\alpha \bullet \beta$ causes the network to withdraw γ, i.e., no longer represent this schema, since the activity in x_4 inhibits x_3. In formal terms $\alpha \bullet \beta\!\mid\!\not\sim\!\gamma$.

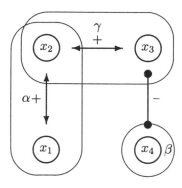

One way of characterizing the nonmonotonic inferences generated by a neural network is to study them in terms of general *postulates* for nonmonotonic logics that have been investigated in the literature (Gabbay [16], Makinson [27], Kraus, Lehmann and Magidor [26], Makinson and Gärdenfors [28], Gärdenfors and Makinson [21]).

It follows immediately from the definition of $[.]\alpha$ that $\mid\!\sim$ satisfies the property of Reflexivity:

$\alpha\!\mid\!\sim\!\alpha$

If we say that a schema β follows logically from α, in symbols $\alpha \vdash \beta$, just when $\alpha \geq \beta$, then it is also trivial to verify that $\mid\!\sim$ satisfies *Supraclassicality*:

If $\alpha \vdash \beta$, then $\alpha\!\mid\!\sim\!\beta$

In words, this property means that immediate consequences of a schema are also nonmonotonic consequences of the schema.

If we turn to the operations on schemata, the following postulate for conjunction is also trivial:

If $\alpha\!\mid\!\sim\!\beta$ and $\alpha\!\mid\!\sim\!\gamma$, then $\alpha\!\mid\!\sim\!\beta \bullet g$ *(And)*

More interesting are the following two properties:

If $\alpha \mathrel{|\!\sim} \beta$ and $\alpha \bullet \beta \mathrel{|\!\sim} \gamma$, then $\alpha \mathrel{|\!\sim} \gamma$ *(Cut)*

If $\alpha \mathrel{|\!\sim} \beta$ and $\alpha \mathrel{|\!\sim} \gamma$, then $\alpha \bullet \beta \mathrel{|\!\sim} \gamma$ *(Cautious Monotony)*

Together Cut and Cautious Monotony are equivalent to each of the following postulates:

If $\alpha \mathrel{|\!\sim} \beta$ and $\beta \vdash \alpha$, then $\alpha \mathrel{|\!\sim} \gamma$ iff $\beta \mathrel{|\!\sim} \gamma$ *(Cumulativity)*

If $\alpha \mathrel{|\!\sim} \beta$ and $\beta \mathrel{|\!\sim} \alpha$, then $\alpha \mathrel{|\!\sim} \gamma$ iff $\beta \mathrel{|\!\sim} \gamma$ *(Reciprocity)*

Cumulativity has become an important touchstone for nonmonotonic systems (Gabbay [16],Makinson [27]). It is therefore interesting to see that the inference operation defined here seems to satisfy Cumulativity (and thus Reciprocity) for almost all neural networks where it is defined. However, it is possible to find some cases where it is not satisfied:

Counterexample to Reciprocity: The network illustrated below is a simple example of a network that does not satisfy Reciprocity (or Cumulativity). For this network it is assumed that all inputs to a neuron are simply added. If we assume that there is a strong excitatory connection between α and β, it follows that $\alpha \mathrel{|\!\sim} \beta$ and $\beta \mathrel{|\!\sim} \alpha$ since α and β do not receive any inhibitory inputs. Suppose that $\alpha = \langle 1\,0\,0\,0 \rangle$ is given as input. From the assumption that the inputs to x_3 interact *additively* it follows that γ receives a larger input than δ, because of the time delay before δ gets activated. If the inhibitory connection between γ and δ is large, the excitatory input from β can never effect the activity of x_3. We then have $\alpha \mathrel{|\!\sim} \gamma$ and $\alpha \mathrel{|\!\not\sim} \delta$. If instead $\beta = \langle 0\,1\,0\,0 \rangle$ is given as input, the situation is the opposite, and so δ gets excited but not γ, and consequently $\alpha \mathrel{|\!\not\sim} \gamma$ and $\alpha \mathrel{|\!\sim} \delta$ Thus, the network does not satisfy Reciprocity.

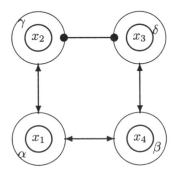

A critical factor here seems to be the *linear* summation of inputs that 'locks' x_2 and x_3 to inputs from the outside because the inhibitory connection between them is large.

Extensive computer simulations have been performed with networks that obey 'shunting' rather than linear summation of excitatory and inhibitory inputs (see Balkenius and Gärdenfors [2]. They suggest that reciprocity is satisfied in all networks that obey this kind of interaction of the inputs.

Shunting interaction of inputs is used in many biologically inspired neural network models (e.g. Cohen and Grossberg [8], Grossberg [24]) and is an approximation of the membrane equations of neurons. A simple example of such a network can be described by the following equation:

$$x_i(t+1) = x_i(t) + \delta(1 - x_i(t)) \sum_j d(x_i(t)) c_{ji}^+ + \delta x_i(t) \sum_j d(x_i(t)) c_{ji}^-$$

Here δ is a small constant, c_{ij}^+ and c_{ij}^- are matrices with all $c_{ij}^+ = c_{ji}^+ \geq 0$, and all $c_{ij}^- = c_{ji}^- \leq 0$; $d(x) \geq 0$ and $d'(x) > 0$. The positive inputs to neuron x_i are shunted by the term $(1 - x_i(t))$ and the negative inputs by $x_i(t)$. As a consequence, the situation where one input locks another of opposite sign cannot occur, in contrast to the linear case above. In other words, a change of input, that is a change in $\sum_j d(x_i(t)) c_{ji}^+$ or $\sum_j d(x_i(t)) c_{ji}^-$, will always change the equilibrium of x_i. The fact that one input never locks another of opposite sign seems to be the reason why all the simulated shunting networks satisfy Reciprocity.

For the disjunction operation it does not seem possible to show that any genuinely new postulates are fulfilled. The following special form of transitivity is a consequence of Cumulativity (cf. Kraus, Lehmann and Magidor [26, p. 179]):

If $\alpha \oplus \beta \hspace{-0.2em}\sim\hspace{-0.2em} \alpha$ and $\alpha \hspace{-0.2em}\sim\hspace{-0.2em} \gamma$, then $\alpha \oplus \beta \hspace{-0.2em}\sim\hspace{-0.2em} \gamma$

This principle is thus satisfied whenever Cumulativity is.

The general form of Transitivity, i.e., if $\alpha \hspace{-0.2em}\sim\hspace{-0.2em} \beta$ and $\beta \hspace{-0.2em}\sim\hspace{-0.2em} \gamma$, then $\alpha \hspace{-0.2em}\sim\hspace{-0.2em} \gamma$, is not valid for all α, β, and γ, as can be shown by the first example above. Nor is Or generally valid:

If $\alpha \hspace{-0.2em}\sim\hspace{-0.2em} \gamma$ and $\beta \hspace{-0.2em}\sim\hspace{-0.2em} \gamma$, then $\alpha \oplus \beta \hspace{-0.2em}\sim\hspace{-0.2em} \gamma$ *(Or)*

Counterexample to Or. The following network is a simple counterexample: x_1 excites x_4 more than x_2 inhibits x_4. The same is true for x_3 and x_2. Giving $\alpha = \langle 1\,1\,0\,0 \rangle$ or $\beta = \langle 0\,1\,1\,0 \rangle$ as input activates x_4, thus $\alpha \hspace{-0.2em}\sim\hspace{-0.2em} \gamma$ and $\beta \hspace{-0.2em}\sim\hspace{-0.2em} \gamma$. On the other hand, the neuron x_2 which represents schema $\alpha \oplus \beta$ has only inhibitory connections to x_4. As a consequence $\alpha \oplus \beta \hspace{-0.2em}\not\sim\hspace{-0.2em} \gamma$.

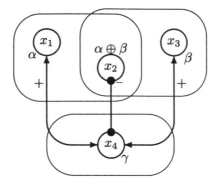

In summary, following Balkenius and Gärdenfors [2], it has been shown that by introducing an appropriate schema concept and exploiting the higher-level features of a resonance function in a neural network it is possible to define a form of nonmonotonic inference relation. It has also been established that this inference relation satisfies some of the most fundamental postulates for nonmonotonic logics.

The construction presented in this section is thus an example of how symbolic features can emerge from the subsymbolic level of a neural network. However, the notion of inference presented here is clearly part of the associationist tradition, since it is based on other primitive notions than what is common within the symbolic paradigm. Sellars [33, p. 265] discusses six conceptions of the status of material rules of inference. The notion presented here fits well with his sixth:

Trains of thought which are said to be governed by "material rules of inference" are actually not inferences at all, but rather activated associations which mimic inference, concealing their intellectual nudity with stolen "therefores."

4.5 Symbolic and Subsymbolic: Two Perspectives on the Same Process

With the two examples from Sections 4.3 and 4.4 in mind, I now want to turn to the current discussion within cognitive science concerning the relation between symbolic and subsymbolic processes. It should be clear that what is normally considered to be a subsymbolic process fits well with the associationist paradigm outlined in Section 4.2. The main thesis of this article is that the symbolic and subsymbolic approaches are not two rival paradigms of computing, rather they are best viewed as two different perspectives that can be adopted when describing the activities of various computational devices.[15] Smolensky [36] formulates the same idea as follows:

"Rather than having to model the mind as *either* a /symbolic/ structure cruncher *or* a number cruncher, we can now see it as a number cruncher in which the numbers crunched are in fact representing complex /symbolic/ structures." (pp. 215–216)

"The connectionist cognitive architecture is intrinsically two-level: Semantic interpretation is carried out at the level of patterns of activity while the complete, precise, and formal account of mental processing must be carried out at the level of individual activity values and connections. Mental processes reside at a lower level of analysis than mental representations." (p. 223)

In the light of this thesis let us look at some of the major discussions concerning the relation between symbolic processing and connectionism, namely, Smolensky [35, 36], Fodor and Pylyshyn [15], Fodor and McLauglin [14].

4.5.1   What is the Proper Treatment of Connectionism?

Smolensky's [35] 'proper treatment of connectionism' (PTC) seems quite closely related to my position. When he argues that the symbolic and the subsymbolic paradigms are incompatible, as I outlined in Section 4.2.3, my interpretation is that he says that the two perspectives cannot be adopted to one level only. As we shall see below, this seems to be exactly what Fodor and Pylyshyn [15] try to do.

There are several aspects of Smolensky's analysis that are similar to the one presented here. For instance, at the end of the article he describes his position as 'emergentist'.[16]

And, as mentioned in Section 4.4.1, his analysis of schemata is congenial with the one presented there. Furthermore, at a first glance at least, his description of inference on the subsymbolic level seems to be quite similar to the account presented in Section 4.4.2:

---

[15]In Gärdenfors [20] I argue that in order to understand inductive reasoning, and thereby cognition in general, one must distinguish between at least three levels: the symbolic (there called linguistic), the conceptual, and the subconceptual level. The distinction between the conceptual and the subconceptual levels is not important for the purposes of the present chapter; they can both be seen as subsymbolic. (However, as pointed out in Gärdenfors [20], Smolensky [35] confuses the symbolic and the conceptual levels).

[16]Also cf. Woodfield and Morton's [40] commentary and the Author's Response on p. 64. In Smolensky [36, p. 202] he writes:

"In giving up symbolic computation to undertake connectionist modeling, we connectionists have taken out an enormous loan, on which we are still paying nearly all interest: solving the basic problems we have created for ourselves rather than solving the problems of cognition. In my view the loan is worth taking out for the goal of understanding how symbolic computation, or approximations of it, can emerge from numerical computation in a class of dynamical systems sharing the most general characteristics of neural computation."

"A natural way to look at the knowledge stored in connections is to view each connection as a *soft constraint*. . . . Formalizing knowledge in soft constraints rather than hard rules has important consequences. Hard constraints have consequences singly; they are rules that can be applied separately and sequentially - the operation of each proceeding independently of whatever other rules may exist. But soft constraints have no implications singly; any one can be overridden by the others. It is only the entire set of soft constraints that has any implications. Inference must be a cooperative process, like the parallel relaxation processes typically found in subsymbolic systems. Furthermore, adding additional soft constraints can repeal conclusions that were formerly valid: Subsymbolic inference is fundamentally nonmonotonic" (1988, p. 18).

However, one worry I have with this description of the activities of a connectionist system is that Smolensky still sees it as performing *inferences* even on the subconceptual level.[17]

This point is made very clearly in Dellarosa's [9, p. 29] commentary:

"It is a belief of many cognitive scientists (most notably, Fodor 1975) that the fundamental process of cognition is inference, a process to which symbolic modelling is particularly well suited. While Smolensky points out that statistical inference replaces logical inference in connectionist systems, he too continues to place inference at the heart of all cognitive activity. I believe that something more fundamental is taking place. In most connectionist models, the fundamental process of cognition is not inference, but is instead the (dear to the heart of psychologists) activation of associated units in a network. Inference 'emerges' as a system-level interpretation of this microlevel activity, but – when representations are distributed – no simple one-to-one mapping of activity patterns to symbols and inferences can be made. From this viewpoint, the fundamental process of cognition is the activation of associated units, and inference is a second-order process."

Thus Smolensky is wrong in talking about 'nonmonotonic inferences' on the subsymbolic level, since there are no inferences on this level; claiming this is basically a kind of category error. However, as has been argued in the previous section, he is right in that the inferences that emerge from the subsymbolic processes on the symbolic level are fundamentally nonmonotonic.

It should be noted that two perspectives on computing that are discussed here are not only applicable to neural networks. Also the behavior of a traditional computer with a von Neumann architecture can be given a 'subsymbolic' interpretation and need

---

[17]Fodor and Pylyshyn [15, pp.29-30] too, make inferences the engine of cognition:

"It would not be unreasonable to describe Classical Cognitive Science as an extended attempt to apply the methods of proof theory to the modeling of thought (and similarly, of whatever mental processes are plausibly viewed as involving inferences; preeminently learning and perception)."

not be seen as merely symbol crunching. The subsymbolic perspective is adopted when one describes the general properties of the physical processes driving the computer; for example when describing the electric properties of transistors. This is the perspective that one must adopt when the computer is defective, in which case the processing on the symbolic level does not function as expected.[18]

A consequence of the fact that one can adopt two perspectives on all kinds of computing devices is that every ascription of symbolic processing to some system is an *interpretation* of the subsymbolic activities. The Turing paradigm of computation neglects this distinction since a computer is thought to uniquely identify some Turing machine; and Turing machines are clearly described on the symbolic level.[19] The reason this identification works is that traditional computers are constructed to be 'digital', i.e., on the subsymbolic perspective the outcomes of the electronic processes are very robust with respect to disturbances so that particular currents can be identified as either '1's or '0's. However, the identification may break down as soon as the computer is malfunctioning.

It follows that the notion of 'computation' can be given two meanings. The first, and to many the only meaning is computation on the symbolic level in the sense that is made precise by 'Turing computable'. According to Church's thesis this kind of computation is all there is *on the symbolic level.* The other sense of computation only becomes apparent when one adopts a subsymbolic (connectionist or more general associationist) perspective. From this perspective 'computation' means 'processing representations', where the representations have a fundamentally different structure compared to those on the symbolic level. And processing on this level does not mean 'manipulating symbols', but must be characterized in other ways. Some kinds of processing of representations on the subsymbolic level generate structures that can be interpreted meaningfully on the symbolic level. However, there are also many kinds of processes that cannot be interpreted on the symbolic level as performing any form of Turing computation. For instance, the notion of 'analog' computation only makes sense on the subsymbolic level. Hence, the class of computational processes on the subsymbolic level is much wider than the class of processes corresponding to Turing computations. Thus, Church's thesis does not apply to this sense of 'computation'.

---

[18]In this context, the subsymbolic perspective is related to adopting the 'physical stance' in the terminology of Dennett [10], while the symbolic level then corresponds to the 'design stance'. The analogy is not perfect since the subsymbolic perspective on the function of a computer need not be tied to a particular physical realization, but can be kept at the level of general functional properties of e.g., transistors, independently of what material they are made from. The same argument, of course, applies to neural networks, the subsymbolic level of which can be described independently of their physical level. The upshot is that the subsymbolic perspective falls 'between' the design stance and the physical stance.
[19]Cf. the quotation from Fodor [13] in Section 4.2.1.

4.5.2   The Compatibility of Symbolism and Connectionism

Let me finally return to Fodor and Pylyshyn's [15] argument against the systematicity and compositionality of connectionism. Their main conclusion is that since cognition is compositional and systematic and since connectionist systems lack those properties, while 'Classical', i.e., symbolic, systems have them, it is only symbolic systems that can represent cognitive processes.

First of all, it should be noted that they assume that, even if there are several levels of analysis, all 'cognitive' levels are representational:

"Since Classicists and Connectionists are both Representationalists, for them any level at which states of the system are taken to encode properties of the world counts as a cognitive level; and no other levels do." (Fodor and Pylyshyn [15, p. 9])

According to them, this assumption about a unique representational level puts a strait-jacket on connectionist methodology:

"It is, for example, *no use at all*, from the cognitive psychologist's point of view, to show that the *non*representational (e.g. neurological, or molecular, or quantum mechanical) states of an organism constitute a Connectionist network, because that would *leave open* the the question whether the mind is such a network *at the psychological level*. It is, in particular, perfectly possible that nonrepresentational neurological states are interconnected in the ways described by Connectionist models *but that the representational states themselves are not*" (p. 10).

So, the key question becomes: How do connectionist systems represent? Fodor and Pylyshyn summarize the disparity between symbolic ('Classical') and connectionist systems as follows:

"Classical and Connectionist theories disagree about the nature of mental representation; for the former, but not for the latter, mental representations characteristically exhibit a combinatorial constituent structure and a combinatorial semantics. Classical and Connectionist theories also disagree about the nature of mental processes; for the former, but not for the latter, mental processes are characteristically sensitive to the combinatorial structure of the representations on which they operate" (Fodor and Pylyshyn [15, p. 32]).

Their main argument for why connectionist systems exhibit neither compositionality (i.e., combinatorial constituent structure) nor systematicity (i.e., sensitivity to this combinatorial structure in processes) is based on their interpretation of how networks represent. It is at this point that they seem to be confusing the symbolic and the connectionist (associationist) perspectives. On p. eq 12 they state that "[r]oughly, Connectionists assign semantic content to 'nodes' [neurons] ...– i.e., to the sorts of things that are typically labeled in Connectionist diagrams; whereas Classicists assign semantic content to expressions – i.e., to the sort of things that get written on the tapes of Turing machines and stored at addresses in von Neumann machines."

Fodor and Pylyshyn's paradigm example of such an assignment is presented on pp. 15-16, where they consider the difference between a connectionist machine handling an inference from $A\&B$ to A and a symbolic machine doing the same thing. They assume that the connectionist machine consists of a network of "labelled nodes" that looks as follows (their Figure 2):

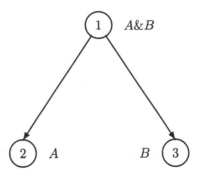

In this network "[d]rawing an inference from $A\&B$ to A thus corresponds to an excitation of node 2 being caused by an excitation of node 1" (p. eq 15).[20] The fundamental mistake in this example is that disjoint nodes are assumed to *represent* different expressions.[21] This assumption conflates representations on the symbolic level (which is where representations of *expressions* and *inferences* belong) with representations on the connectionist level (where representations of associations are handled). Smolensky [36, p. 206] argues concerning Fodor and Pylyshyn's example that

---

[20] A closely related example is given on pp. 47-48 in their paper.
[21] They provide a similar argument on p. 49: "What is deeply wrong with Connectionist architecture is this: Because it acknowledges neither syntactic nor semantic structure in mental representations, it perforce treats them not as a generated set but as a list." However, in fairness it should be acknowledged that Fodor and Pylyshyn [15, p. 12 footnote 7] consider the possibility of having aggregates of neurons representing expressions: "But a subtler reading of Connectionist machines might take it to be the total machine *states* that have content, e.g. the state of *having such and such a node excited*." Even though this comes close to the schema representation presented in Section 4.4.1, they also claim that "[m]ost of the time the distinction between these two ways of talking does not matter for our purposes" [ibid.]. It certainly does, as shall be argued shortly.

"it is a serious mistake to view this as the paradigmatic connectionist account for anything like human inferences of this sort. The kind of *ultralocal* connectionist representation, in which entire propositions are represented by individual nodes, is far from typical of connectionist models, and certainly not to be taken as *definitive* of the connectionist approach."

Given this assumption, it is no wonder that Fodor and Pylyshyn can then argue that networks don't exhibit compositionality and systematicity.

The best way to rebut their argument is to provide a constructive counterexample, i.e., an example of a connectionist system representing things with a compositional and systematic structure. This is readily available from the account of representation and inference in neural networks presented in Section 4.4. There it is *schemata* that represent. Schemata pick out certain *patterns of activities* in the nodes of a connectionist system. As is shown in Section 4.4.2, it is trivial to define some elementary operations on schemata. These operations immediately endow a *compositional* structure on schemata and the components of schemata can be related and combined in a *systematic* way (unlike the one-node representations in Fodor and Pylyshyn's examples).[22] For example, if the schemata $\alpha\bullet\beta$ and $\gamma\oplus\delta$ are both represented in a particular state of a network, one can meaningfully ask whether schemata like $\alpha\bullet\delta$ or $\gamma\oplus\beta$ are also represented in that state (the latter always is). Admittedly, the compositional structure is not very spectacular from a cognitive point of view, but what more can be expected from such a simplistic construction?[23] Furthermore, what is at stake here is not the richness of the representations, but the mere possibility of endowing connectionist systems with a compositional structure of representations.

To be sure, the schemata do not have an *explicit* symbolic structure in the sense that somewhere in the network one finds expressions referring to the schemata (or something representing such expressions). Fodor and Pylyshyn [15, p. 33] seem to think that any productive representational system, i.e., a finitely generated system capable of representing an infinite number of object, must be a symbol system (cf. Bernsen and Ulbæk [5, 6]). However, according to the definition of schemata given in Section 4.4.1, a network with a finite number of nodes can *implicitly* represent an infinite number of schemata.[24] And

---

[22]Cf. Smolensky [36, p. 211]: "Thus in the distributed case, the relation between the node of /the figure above/ labeled $A\&B$ and the others is one kind of whole/part relation. An inference mechanism that takes as input the vector representing $A\&B$ and produces as output the vector representing A is a mechanism that extracts a part from a whole. And in this sense it is no different from a symbolic inference mechanism that takes the syntactic structure **A & B** and extracts from it the syntactic constituent **A**. The connectionist mechanisms for doing this are of course quite different than the symbolic mechanisms, and the approximate nature of the whole/part relation gives the connectionist computation different overall characteristics: we don't have simply a new implementation of the old computation."

[23]Another type of example is provided by Bernsen and Ulbæk [5, 6], who deal with systematicity in representations of spatial relations.

[24]If the neurons can only take a finite number of activity levels, the *references* of the schemata, described

that is sufficient to establish the productivity of this kind of representation. A similar point is made by Smolensky in his [36].

Fodor and McLaughlin [14] have challenged the proposal that vectorial representations in connectionist systems can exhibit systematicity and productivity. The gist of their argument seems to be the following (p. 200):

"...the components of tensor product and superposition vectors differ from Classical constituents in the following way: when a complex Classical symbol is tokened, its constituents are tokened. When a tensor product vector or superposition vector is tokened, its components are not (except per accidens). The implication of this difference, from the point of view of the theory of mental processes, is that whereas the Classical constituents of a complex symbol are, ipso facto, available to contribute to the causal consequences of its tokenings – in particular, they are available to provide domains for mental processes – the components of tensor product and superposition vectors can have no causal status as such. What is merely imaginary can't make things happen, to put this point in a nutshell."

However, the notion of causality Fodor and McLaughlin presumes here is very odd, to say the least – they assume that 'tokenings' of symbols completely decide the causal structure of the mental processes. The subsymbolic processes can, according to them, have no causal role since they are not tokened. To me this seems like saying that the tokenings of '123', '×', '45', and '=' on a pocket calculator are the only causes of a tokening of '5535' appearing in the window, while the underlying electronic processes, not being tokened, can play no causal role.

On the contrary, if we want to analyse the causality of mental processes, we should focus on the subsymbolic level or even the underlying physical processes, while the emerging symbolic structures will, in themselves, not be causally efficacious.[25] Consequently, I believe that Fodor and McLaughlin's attempt to save the argument that the connectionist

---

as vectors of activities, will only constitute a finite class. However, the same applies, for example, to classical propositional logic generated from a finite number of atomic sentences: Even if the language contains an infinite number of formulas, their references, i.e., the propositions expressed (i.e., truth-functions), are finite in number.

[25]Smolensky [36, pp. 222–23] makes this point in the following way: "The Classical strategy for explaining the systematicity of thought is to hypothesize that there is a precise formal account of the cognitive architecture in which the constituents of mental representations have causally efficacious roles in the mental processes acting on them. The PTC view denies that such an account of the cognitive architecture exists, and hypothesizes instead that, like the constituents of structures in quantum mechanics, the systematic effects observed in the processing of mental representations arises because the evolution of vectors can be (at least partially and approximately) explained in terms of the evolution of their components, even though the precise dynamical equations apply at the lower level of the individual numbers comprising the vectors and cannot be pulled up to provide a precise temporal account of the processing at the level of entire constituents – i.e., even though the constituents are not causally efficacious."

approach is not a viable explanation of mental processes is a dead end.

In summary, Fodor and Pylyshyn (and McLaughlin) have put blinders on themselves by only considering a special type of representations in connectionist systems. Given the ensuing narrow field of vision, they can argue that connectionist systems cannot represent what is required for modelling cognition. However, I have argued that once one is allowed to view a wider class of representational possibilities, like, e.g., the schemata of Section 4.4.1, the limitations they point out are no longer there (this is not to say that there are no limitations).

Acknowledgements

I wish to thank Christian Balkenius, Johan van Benthem, Linus Broström, Jan van Eijck, Kenneth Holmqvist, and the Cognitive Science Group in Lund for helpful discussions. My work with this article has been supported by the Swedish Council for Research in the Humanities and Social Sciences.

References

[1] Balkenius, C. (1992): "Neural mechanisms for self-organization of emergent schemata, dynamical schema processing, and semantic constraint satisfaction," manuscript, *Lund University Cognitive Studies* No. 14, Lund University.

[2] Balkenius, C. and P. Gärdenfors (1991): "Nonmonotonic inferences in neural networks," pp. 32-39 in *Principles of Knowledge Representation and Reasoning: Proceedings of the Second International Conference*, J.A. Allen, R. Fikes, and E. Sandewall, eds. (San Mateo, CA: Morgan Kaufmann).

[3] Beale, R. and T. Jackson (1990): *Neural Computing: An Introduction*, Bristol: Adam Hilger.

[4] Benthem, J. van (1992): "Logic and the flow of information," to appear in the *Proceedings of the 9th International Congress of Logic, Methodology, and Philosophy of Science*, Amsterdam: North-Holland.

[5] Bernsen, N. O. and I. Ulbæk (1992): "Two games in Town: Systematicity in distributed connectionist systems", *AISBQ Special Issue on Hybrid Models of Cognition* Part 2, No. 79, 25-30.

[6] Bernsen, N. O. and I. Ulbæk (1992): "Systematicity, thought and attention in a distributed connectionist system," manuscript, Centre of Cognitive Science, Roskilde University.

[7] Churchland, P. S. (1986): *Neurophilosophy: Toward a Unified Science of the Mind/Brain*, Cambridge, MA: MIT Press.

[8] Cohen, M. A. and S. Grossberg (1983): "Absolute stability of global pattern formation and parallel memory storage by competitive neural networks," *IEEE Transactions on Systems, Man, and Cybernetics*, SMC- 13, 815-826.

[9] Dellarosa, D. (1988): "The psychological appeal of connectionism," *Behavioral and Brain Sciences* 11:1, 28-29.

[10] Dennett, D. (1978): Brainstorms, Cambridge, MA: MIT Press.

[11] Eijck, J. van, and F.- J. de Vries (1992): "Dynamic interpretation and Hoare deduction," *Journal of Logic, Language and Information* 1, 1-44.

[12] Fodor, J. (1975): *The Language of Thought*, Cambridge, MA: Harvard University Press.

[13]   Fodor, J. (1981): *Representations*, Cambridge, MA: MIT Press.

[14]   Fodor, J. and B. P. McLaughlin (1990): "Connectionism and the problem of systematicity: Why Smolensky's solution doesn't work," *Cognition* 35, 183-204.

[15]   Fodor, J. and Z. Pylyshyn (1988): "Connectionism and cognitive architecture: A critical analysis," in. S. Pinker & J. Mehler (eds.) *Connections and Symbols*. Cambridge, MA: MIT Press.

[16]   Gabbay, D. (1985): "Theoretical foundations for non-monotonic reasoning in expert systems," in *Logic and Models of Concurrent Systems*, K. Apt ed., Berlin: Springer-Verlag.

[17]   Gärdenfors P. (1984): "The dynamics of belief as a basis for logic," *British Journal for the Philosophy of Science* 35, 1-10.

[18]   Gärdenfors, P. (1985): "Propositional logic based on the dynamics of belief," *Journal of Symbolic Logic* 50, 390-394.

[19]   Gärdenfors, P. (1988): *Knowledge in Flux: Modeling the Dynamics of Epistemic States*, Cambridge, MA: The MIT Press, Bradford Books.

[20]   Gärdenfors, P. (1992): "Three levels of inductive inference," to appear in the *Proceedings of the 9th International Congress of Logic, Methodology, and Philosophy of Science*, Amsterdam: North-Holland.

[21]   Gärdenfors, P. and D. Makinson (1993): "Nonmonotonic inferences based on expectations," to appear in *Artificial Intelligence*.

[22]   Groenendijk, J. and M. Stokhof (1990): "Dynamic Montague grammar," to appear in *Proceedings of the Second Symposium on Logic and Language*, Hajduszoboszlo, Hungary.

[23]   Groenendijk, J. and M. Stokhof (1991): "Dynamic predicate logic," *Linguistics and Philosophy* 14, 39-100.

[24]   Grossberg, S. (1989): "Nonlinear neural networks: Principles, mechanisms, and architectures," *Neural Networks* 1, 17-66.

[25]   Kamp, H. (1981): "A Theory of Truth and Semantic Representation", in Groenendijk, Janssen, Stokhof (eds.), *Formal Methods in the Study of Language*, Mathematisch Centrum, Amsterdam.

[26]   Kraus, S., D. Lehmann, and M. Magidor (1991): "Nonmonotonic reasoning, preferential models and cumulative logics," *Artificial Intelligence* 44, 167-207.

[27]   Makinson, D. (1993): "General patterns in nonmonotonic reasoning," to appear as Chapter 2 of *Handbook of Logic in Artificial Intelligence and Logic Programming, Volume II: Non-Monotonic and Uncertain Reasoning*. Oxford: Oxford University Press.

[28]   Makinson, D. and P. Gärdenfors (1991): "Relations between the logic of theory change and non-monotonic logic," pp. 185-205 in *The Logic of Theory Change*, ed. by A. Fuhrmann and M. Morreau, Springer-Verlag. Also pp. 7-27 in *Proceedings of the Workshop on Nonmonotonic Reasoning*, GMD 1990 (Arbeitspapiere der GMD 443), ed. by G. Brewka and H. Freitag, Gesellschaft für Mathematik und Datenverarbeitung MBH, 1990.

[29]   Pylyshyn, Z. (1984): *Computation and Cognition*, Cambridge, MA: MIT Press.

[30]   de Rijke, M. (1993): "Meeting some neighbours", this volume.

[31]   Rumelhart, D. E. and J. L. McClelland (1986): *Parallel Distributed Processing: Explorations in the Microstructure of Cognition*, Cambridge, MA: MIT Press.

[32]   Rumelhart, D. E., P. Smolensky, J. L. McClelland and G. E. Hinton (1986): "Schemata and sequential thought processes in PDP models," in *Rumelhart, D. E., Parallel Distributed Processing*, Vol. 2, pp. 7-57, Cambridge, MA: MIT Press.

[33]   Sellars, W. (1980): "Inference and meaning," in J. Sicha ed. *Pure Pragmatics and Possible Worlds*, Reseda, CA: Ridgeview Publishing Co.

[34] Smolensky, P. (1986): "Information processing in dynamical systems: foundations of harmony theory," in Rumelhart, D.E., *Parallel Distributed Processing*, Vol. 1, 194-281, Cambridge, MA: MIT Press.

[35] Smolensky, P. (1988): "On the proper treatment of connectionism," *Behavioral and Brain Sciences* 11, 1-23.

[36] Smolenskty, P. (1991): "Connectionism, constituency, and the language of thought," in Loewer, B. and Rey G. (eds.), *Meaning in Mind: Fodor and His Critics*, Oxford: Blackwell, 201-227.

[37] Smolensky, P. (1991): "Tensor product variable binding and the representation of symbolic structures in connectionist systems," *Artificial Intelligence* 46, 159-216.

[38] Thom, R. (1972): *Stabilité Structurelle et Morphogenèse*, New York: Benjamin.

[39] Veltman, F. (1991): "Defaults in update semantics," in M. Moens (ed), *Common sense entailment and update semantics*, DYANA-deliverable R2.5.C. Edinburgh, 1991, 2-60, to appear in *Journal of Philosophical Logic*.

[40] Woodfield, A. and Morton, A. (1988): "The reality of the symbolic and subsymbolic systems," *Behavioral and Brain Sciences* 11, 58.

[41] Zornetzer, S. F., J. L Davis and C. Lau (1990): *An Introduction to Neural and Electronic Networks*, San Diego: Academic Press.

5 On Action Algebras

Dexter Kozen

5.1 Introduction

Action algebras have been proposed by Pratt [22] as an alternative to Kleene algebras [8, 9]. Their chief advantage over Kleene algebras is that they form a finitely-based equational variety, so the essential properties of * (iteration) are captured purely equationally. However, unlike Kleene algebras, they are not closed under the formation of matrices, which renders them inapplicable in certain constructions in automata theory and the design and analysis of algorithms.

In this chapter we consider a class of action algebras called *action lattices*. An action lattice is simply an action algebra that forms a lattice under its natural order. Action lattices combine the best features of Kleene algebras and action algebras: like action algebras, they form a finitely-based equational variety; like Kleene algebras, they are closed under the formation of matrices. Moreover, they form the largest subvariety of action algebras for which this is true. All common examples of Kleene algebras appearing in automata theory, logics of programs, relational algebra, and the design and analysis of algorithms are action lattices.

Iteration is an inescapable aspect of computer programs. One finds a bewildering array of formal structures in the literature that handle iteration in various ways. Many of these are based on the algebraic operator *, a construct that originated with Kleene [12] and has since evolved in various directions. Among these one finds Kleene algebras [3, 8, 9], *-continuous Kleene algebras [3, 7, 10], action algebras [22], dynamic algebras [7, 21], and closed semirings (see [1, 17, 10, 15, 6]), all of which axiomatize the essential properties of * in different ways.

The standard relational and language-theoretic models found in automata theory [4, 5, 16, 15, 14], program logic and semantics (see [11] and references therein), and relational algebra [19, 20] are all examples of such algebras. In addition one finds a number of nonstandard examples in the design and analysis of algorithms, among them the so-called min,+ algebras (see [1, 17, 10]) and certain algebras of polygons [6].

Of the three classes of algebras mentioned, the least restrictive is the class of *Kleene algebras*. Kleene algebras have been studied under various definitions by various authors, most notably Conway [3]. We adopt the definition of [8, 9], in which Kleene algebras are axiomatized by a certain finite set of universally quantified equational implications over the regular operators +, ; , *, 0, 1. Thus the class of Kleene algebras forms a finitely-based equational quasivariety. The equational consequences of the Kleene algebra axioms are exactly the regular identities [3, 8, 13]. Thus the family of regular languages over an alphabet Σ forms the free Kleene algebra on free generators Σ.

A central step in the completeness proof of [8] is the demonstration that the family of

$n \times n$ matrices over a Kleene algebra again forms a Kleene algebra. This construction is also useful in several other applications: matrices over the two-element Kleene algebra are used to derive fast algorithms for reflexive transitive closure in directed graphs; matrices over min,+ algebras are used to compute shortest paths in weighted directed graphs; and matrices over the free monoid Σ^* are used to construct regular expressions equivalent to given finite automata (see [1, 17, 10]). Using matrices over an arbitrary Kleene algebra, one can give a single uniform solution from which each of these applications can be derived as a special case.

Besides equations, the axiomatization of Kleene algebras contains the two equational implications

$$ax \leq x \quad \Rightarrow \quad a^*x \leq x \tag{5.1}$$
$$xa \leq x \quad \Rightarrow \quad xa^* \leq x \ . \tag{5.2}$$

It is known that no finite equational axiomatization exists over this signature [23] (although well-behaved infinite equational axiomatizations have been given [13, 2]). Pratt [22] argues that this is due to an inherent nonmonotonicity associated with the * operator. This nonmonitonicity is handled in Kleene algebras with the equational implications (5.1) and (5.2).

In light of the negative result of [23], it is quite surprising that the essential properties of * should be captured purely equationally. Pratt [22] shows that this is possible over an expanded signature. He augments the regular operators with two *residuation operators* $\rightarrow$ and $\leftarrow$, which give a kind of weak left and right inverse to the composition operator ;, and identifies a finite set of equations that entail all the Kleene algebra axioms, including (5.1) and (5.2). The models of these equations are called *action algebras*. The inherent nonmonotonicity associated with * is captured by the residuation operators, each of which is nonmonotonic in one of its arguments. Moreover, all the examples of Kleene algebras mentioned above have naturally defined residuation operations under which they form action algebras. Thus the action algebras form a finitely-based equational variety contained in the quasivariety of Kleene algebras and containing all the examples we are interested in. This is a desirable state of affairs, since one can now reason about * in a purely equational way.

However, one disadvantage of action algebras is that they are not closed under the formation of matrices. In Example 5.1 below we construct an action algebra $\mathcal{U}$ for which the 2×2 matrices over $\mathcal{U}$ do not form an action algebra. Thus one cannot carry out the program of [8] or use action algebras to give a general treatment of the applications mentioned above that require matrices.

In this chapter we show that the situation can be rectified by further augmenting the signature with a meet operator $\cdot$ and imposing lattice axioms, and that this step is unavoidable if closure under the formation of matrices is desired. Specifically, we show that for $n \geq 2$, the family of $n \times n$ matrices over an action algebra A is again an action algebra if and only if A has finite meets under its natural order. An action algebra with this property is called an *action lattice*. Action lattices have a finite equational axiomatization and are closed under the formation of matrices; moreover, they form the largest subvariety of action algebras for which this is true.

In specializing from action algebras to action lattices, we do not lose any of the various models of interest mentioned above. We have thus identified a class that combines the best features of Kleene algebras and action algebras:

- like action algebras, action lattices form a finitely based equational variety;

- like Kleene algebras, the $n \times n$ matrices over an action lattice again form an action lattice;

- all the Kleene algebras that normally arise in applications in logics of programs, automata theory, relational algebra, and the design and analysis of algorithms are examples of action lattices.

5.2 Definitions

With so many operators and axioms, it is not hard to become confused. Not the least problem is conflict of notation in the literature. For the purposes of this chapter, we follow [22] and use $+$ and $\cdot$ for join and meet, respectively, and $;$ for composition ([8, 9, 10] use $\cdot$ for composition).

For ease of reference, we collect all operators, signatures, axioms, and classes of structures together in four tables. All classes of algebraic structures we consider will have signatures consisting of some subset of the operators in Table 5.1 and axioms consisting of some subset of the formulas of Table 5.3. The signatures and classes themselves are defined in Tables 5.2 and 5.4, respectively.

The binary operators are written in infix. We normally omit the operator $;$ from expressions, writing ab for $a; b$. We avoid parentheses by assigning $*$ highest priority, then $;$, then all the other operators. Thus $a + bc^*$ should be parsed $a + (b(c^*))$.

The expression $a \leq b$ is considered an abbreviation for the equation $a + b = b$.

As shown in [22], the two definitions of **RES** given in Table 5.4 are equivalent. The first gives a useful characterization of $\rightarrow$ and $\leftarrow$ in succinct terms, and the second gives a purely equational characterization. With the second definition, **RES** and **ACT** are defined by pure equations.

Let **C** be a class of algebraic structures with signature σ, and let $\mathcal{A}$ be an algebraic structure with signature τ. We say that $\mathcal{A}$ *expands to* an algebra in **C** if the operators in $\sigma - \tau$ can be defined on $\mathcal{A}$ in such a way that the resulting algebra, restricted to signature σ, is in **C**.

symbol	name	arity
$+$	sum, join, plus	2
;	product, (sequential) composition	2
$\cdot$	meet	2
$\leftarrow$	left residuation	2
$\rightarrow$	right residuation	2
$*$	star, iteration	1
0	zero, additive identity	0
1	one, multiplicative identity	0

Table 5.1
Operators.

short name	name	operators
is	idempotent semirings	$+$, ;, 0, 1
ka	Kleene algebras	**is**, $*$
res	residuation algebras	**is**, $\leftarrow$, $\rightarrow$
act	action algebras	**ka**, **res**
al	action lattices	**act**, $\cdot$

Table 5.2
Signatures.

$$a + (b + c) \;=\; (a + b) + c \tag{5.3}$$

$$a + b \;=\; b + a \tag{5.4}$$

$$a + a \;=\; a \tag{5.5}$$

$$a + 0 \;=\; 0 + a \;=\; a \tag{5.6}$$

$$a(bc) \;=\; (ab)c \tag{5.7}$$

$$a1 \;=\; 1a \;=\; a \tag{5.8}$$

$$a(b + c) \;=\; ab + ac \tag{5.9}$$

$$(a + b)c \;=\; ac + bc \tag{5.10}$$

$$a0 \;=\; 0a \;=\; 0 \tag{5.11}$$

$$1 + a + a^*a^* \;\leq\; a^* \tag{5.12}$$

$$ax \leq x \;\Rightarrow\; a^*x \leq x \tag{5.13}$$

$$xa \leq x \;\Rightarrow\; xa^* \leq x \tag{5.14}$$

$$ax \leq b \;\Longleftrightarrow\; x \leq a \rightarrow b \tag{5.15}$$

$$xa \leq b \;\Longleftrightarrow\; x \leq b \leftarrow a \tag{5.16}$$

$$a(a \rightarrow b) \;\leq\; b \tag{5.17}$$

$$(b \leftarrow a)a \;\leq\; b \tag{5.18}$$

$$a \rightarrow b \;\leq\; a \rightarrow (b + c) \tag{5.19}$$

$$b \leftarrow a \;\leq\; (b + c) \leftarrow a \tag{5.20}$$

$$x \;\leq\; a \rightarrow ax \tag{5.21}$$

$$x \;\leq\; xa \leftarrow a \tag{5.22}$$

$$(x \rightarrow x)^* \;=\; x \rightarrow x \tag{5.23}$$

$$(x \leftarrow x)^* \;=\; x \leftarrow x \tag{5.24}$$

$$a \cdot (b \cdot c) \;=\; (a \cdot b) \cdot c \tag{5.25}$$

$$a \cdot b \;=\; b \cdot a \tag{5.26}$$

$$a \cdot a \;=\; a \tag{5.27}$$

$$a + (a \cdot b) \;=\; a \tag{5.28}$$

$$a \cdot (a + b) \;=\; a \tag{5.29}$$

Table 5.3
Axioms.

class	name	sign.	defining axioms
US	upper semilattices	+	(5.3)–(5.5)
IS	idempotent semirings	is	(5.3)–(5.11)
KA	Kleene algebras	ka	**IS**, (5.12)–(5.14)
RES	residuation algebras	res	**IS**, (5.15)–(5.16)
RES	residuation algebras	res	**IS**, (5.17)–(5.22)
RKA	residuated Kleene alg.	act	**KA**, **RES**
ACT	action algebras	act	**RES**, (5.12), (5.23), (5.24)
LS	lower semilattices	·	(5.25)–(5.27)
L	lattices	+, ·	**US**, **LS**, (5.28), (5.29)
AL	action lattices	it	**ACT**, **L**

Table 5.4
Algebraic structures.

5.3   Main Results

5.3.1   Action Algebras are Residuated Kleene Algebras

We first give an alternative characterization of action algebras that we will later find useful: action algebras are exactly the residuated Kleene algebras.

LEMMA 5.1   **ACT = RKA.**

Proof   Every action algebra is a residuation algebra by definition. As shown in [22], every action algebra is a Kleene algebra. This establishes the forward inclusion.

Conversely, we show that the properties (5.23) and (5.24) hold in all residuated Kleene algebras. By symmetry, it will suffice to show (5.23). The inequality $x \to x \le (x \to x)^*$ follows from (5.12) and the **IS** axioms. For the reverse inequality, we have

$$
\begin{aligned}
x(x \to x) &\le x && \text{by (5.17)} \\
x(x \to x)^* &\le x && \text{by (5.14), and} \\
(x \to x)^* &\le x \to x && \text{by (5.15).}
\end{aligned}
$$

□

5.3.2   Matrices

Let $\mathcal{R}$ be an idempotent semiring and let $\mathbf{Mat}(n, \mathcal{R})$ denote the family of $n \times n$ matrices over $\mathcal{R}$, with + interpreted as the usual matrix addition, ; the usual matrix multiplication,

0 the zero matrix, and 1 the identity matrix. Under these definitions, $\mathbf{Mat}(n, \mathcal{R})$ forms an idempotent semiring. Moreover, if $\mathcal{R}$ is also a Kleene algebra, we define * on $\mathbf{Mat}(n, \mathcal{R})$ in the usual way (see [3, 8, 10]); then $\mathbf{Mat}(n, \mathcal{R})$ forms a Kleene algebra [8].

We say that an ordered structure $\mathcal{R}$ *has finite meets* if every finite set of elements has a meet or greatest lower bound. An upper semilattice $(R, +)$ has (nonempty) finite meets if and only if it expands to a lattice $(R, +, \cdot)$; the operation $\cdot$ gives the meet of its arguments.

LEMMA 5.2   Let $\mathcal{R} = (R, +, ;, 0, 1, \leftarrow, \rightarrow)$ be a residuation algebra. For $n \geq 2$, the idempotent semiring $\mathbf{Mat}(n, \mathcal{R})$ expands to a residuation algebra if and only if $\mathcal{R}$ has finite meets.

Proof   Suppose first that $\mathcal{R}$ has finite meets, and expand $\mathcal{R}$ to a lattice $(R, +, \cdot)$ accordingly. Using the notation $\sum$ for iterated $+$ and $\prod$ for iterated $\cdot$, we define the operations $\rightarrow$ and $\leftarrow$ on $\mathbf{Mat}(n, \mathcal{R})$ as follows:

$$(A \rightarrow B)_{ij} \;=\; \prod_{k=1}^{n}(A_{ki} \rightarrow B_{kj}) \tag{5.30}$$

$$(B \leftarrow A)_{ij} \;=\; \prod_{k=1}^{n}(B_{ik} \leftarrow A_{jk}) \tag{5.31}$$

Then for all $n \times n$ matrices X,

$$
\begin{aligned}
AX \leq B \;&\Longleftrightarrow\; \bigwedge_{ij}(AX)_{ij} \leq B_{ij} \\
&\Longleftrightarrow\; \bigwedge_{ij}(\sum_k A_{ik}X_{kj}) \leq B_{ij} \\
&\Longleftrightarrow\; \bigwedge_{ij}\bigwedge_{k} A_{ik}X_{kj} \leq B_{ij} \\
&\Longleftrightarrow\; \bigwedge_{jk}\bigwedge_{i} X_{kj} \leq A_{ik} \rightarrow B_{ij} \\
&\Longleftrightarrow\; \bigwedge_{jk} X_{kj} \leq \prod_{i} A_{ik} \rightarrow B_{ij} \\
&\Longleftrightarrow\; \bigwedge_{jk} X_{kj} \leq (A \rightarrow B)_{kj} \\
&\Longleftrightarrow\; X \leq A \rightarrow B .
\end{aligned}
$$

The property

$$XA \leq B \;\Longleftrightarrow\; X \leq B \leftarrow A$$

follows from a symmetric argument. Thus the residuation axioms (5.15) and (5.16) are satisfied in $\mathbf{Mat}(n, \mathcal{R})$ with these definitions.

Conversely, suppose $\mathbf{Mat}(2, \mathcal{R})$ expands to a residuation algebra (the argument is similar for any $n > 2$). Then with respect to the natural order $\leq$ in $\mathcal{R}$ defined in terms of $+$, there exist maximum x, y, z, w such that

$$\begin{bmatrix} 1 & 0 \\ 1 & 0 \end{bmatrix} \begin{bmatrix} x & y \\ z & w \end{bmatrix} \leq \begin{bmatrix} a & a \\ b & b \end{bmatrix}$$

componentwise; i.e., x, y, z, w are maximum such that $x \leq a$, $x \leq b$, $y \leq a$, and $y \leq b$. Then x and y are the greatest lower bound of a and b with respect to $\leq$. Since a and b were arbitrary, R contains all binary meets, hence all nonempty finite meets. The empty meet is given by the top element $0 \to 0$.   □

Not every residuation algebra has finite meets; we construct a counterexample below. Thus the family of $n \times n$ matrices over a residuation algebra does not in general form a residuation algebra. The same is true for action algebras. Hence, in order to obtain a subvariety of action algebras closed under the formation of matrices, we will be forced to account for $\cdot$ explicitly.

EXAMPLE 5.1   *We construct an action algebra that does not have finite meets. Let $\mathcal{U}$ be an arbitrary upper semilattice containing three elements $0, 1, T$ such that $0 < 1 \leq u \leq T$ for all $u \neq 0$. Let $+$ be the join operation of $\mathcal{U}$, let the distinguished elements 0, 1 be as given, and define the remaining action algebra operations as follows:*

$$ab = ba = \begin{cases} 0 & \text{if } a = 0 \text{ or } b = 0 \\ b & \text{if } a = 1 \\ T & \text{if both } a, b > 1 \end{cases}$$

$$a^* = \begin{cases} 1 & \text{if } a = 0 \text{ or } a = 1 \\ T & \text{if } a > 1 \end{cases}$$

$$a \to b = b \leftarrow a = \begin{cases} 0 & \text{if } a \not\leq b \\ 1 & \text{if } 1 < a \leq b < T \\ b & \text{if } a = 1 \\ T & \text{if } a = 0 \text{ or } b = T \end{cases}$$

It is straightforward to check that the resulting structure is an action algebra. Moreover, $\mathcal{U}$ can certainly be chosen without finite meets; for example, let $\mathcal{U}$ consist of the natural numbers, two incomparable elements above the natural numbers, and a top element. Then the two incomparable elements have no meet.   □

We show now that if $\cdot$ is added to the signature of action algebras along with the lattice equations, we obtain a finitely-based subvariety **AL** of **ACT** closed under the formation of matrices. Moreover, it is the largest subvariety of **ACT** with this property, by the direction ($\leftarrow$) of Lemma 5.2.

THEOREM 5.1   The Kleene algebra $\mathbf{Mat}(n, \mathcal{A})$ of $n \times n$ matrices over an action lattice $\mathcal{A}$ expands to an action lattice.

Proof As remarked previously, $\mathbf{Mat}(n, \mathcal{A})$ forms a Kleene algebra under the usual definitions of the Kleene algebra operations $+$, $;$, *, $,0$, 1 [8]. Let the residuation operations be defined as in (5.30) and (5.31); by Lemma 5.2, $\mathbf{Mat}(n, \mathcal{A})$ is a residuation algebra. Then by Lemma 5.1, $\mathbf{Mat}(n, \mathcal{A})$ is an action algebra. Finally, let $\cdot$ be defined on matrices componentwise. Since $\mathcal{A}$ is a lattice and since $+$ and $\cdot$ are defined componentwise, $\mathbf{Mat}(n, \mathcal{A})$ is also a lattice (it is isomorphic to the direct product of n^2 copies of $\mathcal{A}$). Thus $\mathbf{Mat}(n, \mathcal{A})$ is an action lattice.   □

All the examples given in §5.1, under the natural definitions of the residuation and meet operators, are easily seen to be examples of action lattices. Thus we have given a finitely-based variety **AL** that contains all these natural examples and is closed under the formation of $n \times n$ matrices.

5.4   Conclusions and Open Questions

The Kleene algebras have a natural free model on free generators Σ, namely the regular sets $\mathbf{Reg}_\Sigma$ [8]. This structure expands to an action algebra under the natural definition of the residuation operators

$$A \rightarrow B = \{x \in \Sigma^* \mid \forall y \in A \; yx \in B\}$$
$$B \leftarrow A = \{x \in \Sigma^* \mid \forall y \in A \; xy \in B\}$$

and to an action lattice under the definition

$$A \cdot B = A \cap B.$$

Thus the axioms of action algebras and action lattices do not entail any more identities over the signature **ka** than do the Kleene algebra axioms.

One might suspect from this that $\mathbf{Reg}_\Sigma$ with residuation is the free action algebra on Σ and $\mathbf{Reg}_\Sigma$ with residuation and meet is the free action lattice on Σ, but this is not the case: the identity

$$a \rightarrow (a + ba) = 1$$

holds in $\mathbf{Reg}_{\{a,b\}}$, but is not a consequence of the axioms of action algebras or action lattices, as can be seen by reinterpreting $a \mapsto a$ and $b \mapsto a$.

We conclude with some open questions.

1. What is the complexity of the equational theory of action algebras and action lattices? (The equational theory of Kleene algebras is *PSPACE*-complete [18].)

2. Every *-continuous Kleene algebra extends universally to a closed semiring in the sense that the forgetful functor from closed semirings to *-continuous Kleene algebras has a left adjoint [9]. In a sense, this says that it does not matter which of the two classes one chooses to work with. Is there such a relationship between Kleene algebras and action algebras, or between action algebras and action lattices?

Acknowledgements

These results were obtained at the workshop "Logic and the Flow of Information" held on December 13–15, 1991 in Amsterdam, sponsored by NFI project NF 102/62-356, "Structural and Semantic Parallels in Natural Languages and Programming Languages." I am grateful to the workshop organizers Jan van Eijck and Johan van Benthem for their superb organization and to the proprietors of the Hotel De Filosoof for providing a most congenial atmosphere. I would also like to thank the workshop participants, particularly Larry Moss and Dana Scott, for valuable comments. Finally, I am deeply indebted to Vaughan Pratt for insight and inspiration derived from his paper [22] and from many engaging discussions.

References

[1]   Alfred V. Aho, John E. Hopcroft, and Jeffrey D. Ullman. *The Design and Analysis of Computer Algorithms*. Addison-Wesley, 1975.

[2]   Stephen L. Bloom and Zoltán Ésik, "An equational axiomatization of the regular sets," manuscript, April 1991.

[3]   John Horton Conway. *Regular Algebra and Finite Machines*. Chapman and Hall, London, 1971.

[4]   F. Gécseg and I. Peák. *Algebraic Theory of Automata*. Akadémiai Kiadó, Budapest, 1972.

[5]   John E. Hopcroft and Jeffrey D. Ullman. *Introduction to Automata Theory, Languages, and Computation*. Addison-Wesley, 1979.

[6]   Kazuo Iwano and Kenneth Steiglitz, "A semiring on convex polygons and zero-sum cycle problems," *SIAM J. Comput.* 19:5 (1990), 883–901.

[7]   Dexter Kozen, "On induction vs. *-continuity," *Proc. Workshop on Logics of Programs 1981*, Spring-Verlag Lect. Notes in Comput. Sci. 131, ed. Kozen, 1981, 167-176.

[8]   Dexter Kozen, "A completeness theorem for Kleene algebras and the algebra of regular events," *Proc. 6th IEEE Symp. Logic in Comput. Sci.*, July 1991, 214–225. Submitted, *Information and Computation*.

[9]   Dexter Kozen, "On Kleene algebras and closed semirings," *Proc. Math. Found. Comput. Sci. 1990*, ed. Rovan, Lect. Notes in Comput. Sci. 452, Springer, 1990, 26–47.

[10] Dexter Kozen, *The Design and Analysis of Algorithms*. Springer-Verlag, 1991.

[11] Dexter Kozen and Jurek Tiuryn, "Logics of Programs," in: van Leeuwen (ed.), *Handbook of Theoretical Computer Science*, v. B, North Holland, Amsterdam, 1990, 789–840.

[12] Stephen C. Kleene, "Representation of events in nerve nets and finite automata," in: *Automata Studies*, ed. Shannon and McCarthy, Princeton U. Press, 1956, 3–41.

[13] Daniel Krob, "A complete system of B-rational identities," Technical Report 90-1, Institute Blaise Pascal, Paris, January 1990.

[14] Werner Kuich, "The Kleene and Parikh Theorem in complete semirings," in: *Proc. 14th Colloq. Automata, Languages, and Programming*, ed. Ottmann, Springer-Verlag Lecture Notes in Computer Science 267, 1987, 212–225.

[15] Werner Kuich and Arto Salomaa. *Semirings, Automata, and Languages*. Springer-Verlag, Berlin, 1986.

[16] Harry Lewis and Christos Papadimitriou. *Elements of the Theory of Computation*. Prentice-Hall, 1981.

[17] Kurt Mehlhorn. *Data Structures and Algorithms 2: Graph Algorithms and NP-Completeness*. EATCS Monographs on Theoretical Computer Science, Springer-Verlag, 1984.

[18] Albert R. Meyer and Larry Stockmeyer, "The equivalence problem for regular expressions with squaring requires exponential time," in: *Proc. 13th IEEE Symp. on Switching and Automata Theory*, Long Beach, CA, 1972, 125–129.

[19] K. C. Ng and A. Tarski, "Relation algebras with transitive closure," Abstract 742-02-09, *Notices Amer. Math. Soc.* 24 (1977), A29-A30.

[20] K. C. Ng. *Relation Algebras with Transitive Closure*. PhD Thesis, University of California, Berkeley, 1984.

[21] Vaughan Pratt, "Dynamic algebras as a well-behaved fragment of relation algebras," in: D. Pigozzi, ed., *Proc. Conf. on Algebra and Computer Science*, Ames, Iowa, June 2-4, 1988; Springer-Verlag Lecture Notes in Computer Science, to appear.

[22] Vaughan Pratt, "Action logic and pure induction," in: *Proc. Logics in AI: European Workshop JELIA '90*, ed. J. van Eijck, Springer-Verlag Lect. Notes in Comput. Sci. 478, September 1990, 97–120.

[23] V. N. Redko, "On defining relations for the algebra of regular events," *Ukrain. Mat. Z.* 16 (1964), 120–126 (in Russian).

[24] Arto Salomaa, "Two complete axiom systems for the algebra of regular events," *J. Assoc. Comput. Mach.* 13:1 (January, 1966), 158–169.

[25] Arto Salomaa and Matti Soittola. *Automata Theoretic Aspects of Formal Power Series*. Springer-Verlag, 1978.

6 Logic and Control: How They Determine the Behaviour of Presuppositions

Marcus Kracht

6.1 Three Problems for Presuppositions

Presupposition is one of the most important phenomena of non-classical logic as concerns the applications in philosophy, linguistics and computer science. The literature on presuppositions in linguistics and analytic philosophy is rather rich (see [13] and the references therein), and there have been numerous attempts in philosophical logic to solve problems arising in in connection with presuppositions such as the projection problem. In this essay I will introduce a system of logics with control structure and elucidate the relation between context-change potential, presupposition projection and three-valued logic.

For a definition of what presuppositions are consider these three sentences.

(1) *Hilary is not a bachelor.*
(2) *The present king of France is not bald.*
(3) $\lim_{n \to \infty} a_n \neq 4$

Each of these sentences is negative and yet there is something that we can infer from them as well as from their positive counterparts; namely the following.

(1[†]) Hilary is male.
(2[†]) France has a king.
(3[†]) $(a_n)_{n \in \mathbb{N}}$ is convergent.

This is impossible under classical circumstances. In classical logic, nothing of significance can be inferred from both P and $\neg P$ – but here we can infer non-trivial conclusions from both a sentence and its negation. Exactly how does this come about? The most popular answer has been given by Strawson. According to him a sentence may or may not *assert* something; the conditions under which a sentence asserts are not only syntactic but also *semantic* in nature. So, while *Dog the table very which under* is syntactically ill-formed and for that reason fails to assert, the sample sentences given above are syntactically well-formed and yet may fail to assert, namely when some conditions are not met. According to Strawson we say that a sentence S **presupposes** another sentence T if whenever S asserts, T is true. For example, (3) presupposes a_n *is convergent* since the former is assertive only if the latter is true. We will not question this view here; all we ask of the reader at this stage is his consent that the given intuitions are sound. If they are, sentences can no longer be equated with propositions. A sentence is a proposition only if it is assertive. Assertivity depends on the facts and hence it is not possible to say outright whether a given sentence is a proposition; this can vary from situation to situation. The distinction between sentences and propositions carries over to logic if we want to hold on to the assumption that sentences must have truth values. Then, as the classical truth-values shall continue to function in the same way, in a given model a proposition is still defined to be a sentence that is either true or false. On the grounds that there exist non-propositions we need to postulate at least one more truth-value, which, by penalty of self-contradiction, cannot mean *has no truth-value*; rather, it means *has no classical truth value.*

Three problems of presupposition theory can be isolated with which we will deal in turn. These are the *separation problem*, the *projection problem* and the *allocation problem*. The projection problem has a long intellectual history in linguistics. Intuitions have oscillated between an interpretation of presuppositions as a reflex of logic or as a result from the speaker-hearer-interaction, in short between a purely semantic and a pragmatic account. Strong Russellianists such as [9] want to deny it any status in semantics while most semanticists try to derive as many of the projection phenomena from their theory as they can. We will see shortly that both must be wrong. If semantics has to do with meaning that is static (at least in the short run) then computer languages and mathematical jargon provide solid evidence that three valued logic is here to stay and presupposition has a home in semantics. Yet, if one and the same sentence has two different meanings, that is, if we acknowldege that there are cases of ambiguity with respect to the presuppositions which are only resolved by the context, there must be more to presupposition than a semantical theory can provide. The problem of projection is generally stated as follows. Suppose that a sentence S is built from some simple sentences $S_1, \ldots, S_n$ and that we know the presuppositions of $S_1, \ldots, S_n$, can we compute the presuppositions of S? This is really a non-trivial question. A first inspection of examples suggests that presuppositions are simply accumulated; this is the theory of [10]. But consider (4).

(4)   If $(a_n)_{n \in \mathbb{N}}$ is convergent then $\lim_{n \to \infty} a_n \neq 4$.

It has quickly been found that if S presupposes T then T *and* S as well as *If* T *then* S do not presuppose T. So there is a general question as to why this is so and what other rules of projection are valid. It has gone unnoticed that the way in which we have stated the projection problem it becomes ambiguous or at least hopelessly untractable in view of the data that has been accumulated over the years. It is known, namely, that the logical form of a complex sentence need not directly conform to the logical form of the message communicated. Subordination provides one example; another, rather vexing example is the following announcement that could be seen e. g. in a cinema. (This example is due to [12].)

(5)   Old age persons $\left\{ \begin{array}{c} and \\ or \end{array} \right\}$ students at half price.

No matter whether the board says *and* or *or*, we read the same message out of it. This means that we have to postulate two levels of representation similar to syntax: the surface form, called here *syntactic logical form*, and the logical form of the message that is communicated. We use the word *message* rather loosely here but it should be clear that it is not the same as the utterance of the sentence or meaning thereof. The logical form of the message will be called the *underlying* (or *semantic*) *logical form*. The map from the syntactic logical form to the underlying logical form is not unique and it is not at all clear how the two relate; to spell this out in detail is the the *problem of the underlying logical form*. Once the underlying logical form is found, the question how the presupposition of a complex expression is computed can be asked again and answers can be given by direct calculation; we will show how this is done using three valued logic. It is in the following sense that I will understand the projection problem: Given the underlying logical form of a sentence, what are its presuppositions?

Distinct from the projection problem in the narrow sense is the *allocation problem*.[1] To formulate it we assume that the semantics of words or phrases gives rise to explicit presuppositions. Though there is no surface connective that is equivalent to the presuppositionification operators $\downarrow$ or ∇ to be defined later, there are constructions or words that need to be translated using $\downarrow$ or ∇. Typical examples are *know* or *bachelor*. In DRT terms, they create a special presuppositional box which I call a *semantic anchor*. This anchor needs to be dropped somwhere. That there really is a choice between places at which to drop this anchor, let us consider the next examples.

(6)   Everytime X saw four aces in Y's hand he signalled to *his partner*.

(7)   If the judges make a mistake in the formal procedure *the lawyer* will persuade *his client* to appeal.

In (6), we can assume X to have a fixed partner, that is, we can read into (6) a *stronger statement than the one given*. (I should excuse myself here for not being precise; of course, by *statement given* I mean something like the syntactic logical form but this is not 100% right.) But we need not; if we assume that X is playing in each game with a different partner, the referent of *his partner* will depend on the chosen occasion in which X sees four aces in Y's hand. Similarly with (7). If the presupposition initiated by the phrase *the lawyer* stays local, (7) presupposes (7[†]). If it is chosen to be global, (7) presupposes (7[‡]).

---

[1]I will later challenge the picture on presupposition allocation that I will now draw. For the moment it suffices that the problem itself becomes clear.

($7^\dagger$)   If the judges make a mistake in the formal procedure there is one and only one lawyer.

($7^\ddagger$)   There is a unique lawyer.

Admittedly, ($7^\dagger$) sounds artificial because we are more inclined to say that the lawyer is sufficiently determined by the legal procedure and not the additional formal mistakes that may occur. We will see that this view is not justified; but even if it were this only fortifies our arguments that projection and allocation should be seen in the context of the problem of logical form. To account for these differences, we introduce a distinction between the *origin* of a presupposition and its *locus*. First we fix an underlying logical form with respect to the language, that is, we do not yet interpret the words by a semantical meta- or infra-language; we can assume this level to be some variant of LF in Government and Binding. Once we have done that we unpack the meaning of the occurring words in terms of the representational language, a process which accidentally also produces instances of presupposition creating operators ($\downarrow$ or ∇). Or, to use the DRT metaphor, the anchors are now being dropped. This is a straightforward translation; we call the place in the syntactic parse of the resulting translation at which a presupposition is created the **origin**. Mostly, the origin can be located in the original sentences by pointing at the item creating the presupposition. The representation thus created is only a transient one. For we now consider the question whether the presupposition should be placed somewhere else in order to obtain the correct underlying logical form. This 'placing somewhere else' can be understood as a kind of movement transformation that will remove the presupposition from its origin and reinsert it somewhere else. In order not to offend established associations I refer to this process as *reallocation*. The place at which a presupposition is finally placed is called the **locus**. [14] can be understood as a theory of allocation in our sense.

The **separation problem** originates from a distinction between *the* assertion of a sentence and *the* presupposition of a sentence. It consists in the problem to find, given a sentence S, two propositions (!) A and P such that
(sep) S is true iff A is true
          S is a proposition iff P is true
The separation problem has received little attention; to our knowledge it has never been explicitly formulated. However, it is quite an important one since normal semantic theories assume that all their predicates used in the representation are bivalent. So, when *bachelor* finally receives an interpretation via, say, Montague translation, as $\mathbf{male'}(x) \wedge \mathbf{human'}(x) \wedge \mathbf{adult'}(x) \wedge \mathbf{unmarried'}(x)$ it is assumed that $\mathbf{male'}(x)$, $\mathbf{human'}(x)$, $\mathbf{adult'}(x)$ as well as $\mathbf{unmarried'}(x)$ are classical and can therefore be manipulated on the basis of a distinction between truth and falsity only. Many projection algorithms are defective in the sense that it they tacitly assume that they manipulate

only propositions. In ordinary language, however, it is not at all clear that separation can be fully carried out. For even though we can name *sentences* that fulfill (sep) it is not clear whether we can have *propositions* to fulfill (sep). This is due to the fact that all predicates and operators in language are *typed*, that is, they need as input an object of a certain type in order to be well-formed. For example, **unmarried′**(x) requires an x that is human; otherwise *Consciousness is unmarried* would count as a proposition.

Under limited circumstances, however, separation can be worked out to the bottom. Such circumstances are provided in mathematics. For example, $\lim_{n\to\infty} a_n \neq 4$ can be separated into

A : No density point of $(a_n)_{n\in\mathbb{N}}$ is equal to 4.
P : $(a_n)_{n\in\mathbb{N}}$ is bounded and has exactly one density point.

Likewise, $a/x = 6$ can be separated into

$A : a = 6x$
$P : x \neq 0$

The projection problem may also be formulated as follows. Given a sentence S composed from simple sentences $S_1, \ldots, S_n$; how to separate S on the basis of a separation $A_1 : P_1, \ldots, A_n : P_n$?

6.2   Some Notions from Logic

Propositional languages have the advantage of knowing only one type of well-formed expression, that of a proposition. Since there is a clash with the philosophical terminology I will refer to the propositions of an arbitrary propositional language as *terms*. Terms are produced from variables and connectives in the known way. Terms are interpreted in *algebras*. A propositional language defines a similarity type of algebras in which we can interpret the variables and also the terms. Let us fix such a language and call it $\mathbb{L}$. A *logic* over $\mathbb{L}$ is defined via a set of *rules* in the obvious way. A rule is a pair $\langle \Delta, Q \rangle$ where Δ is a finite set of terms called the *premisses* and Q a single term called the *conclusion*. We will not spell out the details here and refer instead to [17]. Logics correspond one-to-one with certain classes of *matrices*. A **matrix** is a pair $\mathfrak{M} = \langle \mathfrak{A}, D \rangle$ where $\mathfrak{A}$ is an algebra of the similarity type of $\mathbb{L}$ and D a set of elements of $\mathfrak{A}$. D is the set of *designated elements* of $\mathfrak{M}$. The rule $\langle \Delta, Q \rangle$ is *valid* in $\mathfrak{M}$ is for all valuations β we have $\beta(Q) \in D$ if only $\beta(\Delta) \subseteq D$. We write $\Delta \vdash_{\mathfrak{M}} Q$.

It is possible to give an analogical treatment of presupposition. In addition to designated truth values we need a distinction between **admitted** and **non-admitted** or **unwanted** truth-values. Technically, if we want to incorporate presupposition into logic we have to expand logical matrices by a set that tells us which truth-values are unwanted, just as we need a set of designated truth-values to tell us what truth is. It seems natural to say that designated truth-values are admitted and hence we get the following definition.

DEFINITION 6.1   A *p-matrix* or *presuppositional matrix* is a triple $\mathfrak{P} = \langle \mathfrak{A}, D, U \rangle$ where $D \subseteq A$ is a set of designated elements and $U \subseteq A$ a set of unwanted truth values, and moreover $D \cap U = \emptyset$. We say that P *presupposes* Q *relative to* $\mathfrak{P}$ – in symbols $P \rhd_{\mathfrak{P}} Q$ – if for all valuations β $\beta(P) \notin U$ implies $\beta(Q) \in D$.

A general theory of presuppositions in arbitrary languages is possible but we prefer to concentrate on three valued logic in relation to the three main issues of presuppositional theory. This does by no means imply that the abstract approach sketched here serves no real purpose. Indeed, in computer science there are more than one recognizable type of unwanted truth value, namely loop and fail. Moreover, in more sophisticated logics for natural language there sometimes is a need to have more than two truth-values. In all of these cases, a presuppositional theory can be added on top using these abstract methods.

6.3   Connectives with Explicit Control Structure

Unlike classical logic, three-valued logic forces us to think quite seriously about the meaning of simple connectives such as *and* and *or*. Indeed, there is no single best choice of a three-valued interpretation. Rather than arguing for one interpretation that it is best we will try to develop an understanding of the difference between these options. We will use a computational interpretation which has its origin in the discussion of [5]. At the heart of this interpretation lies a consistent reading of the third truth-value as computational failure (fail or, in our context U). This failure arises from improper use of partial predicates or functions, e.g., dividing by 0, taking the square root of negative numbers, etc. A second component is the addition of an explicit *control structure* that determines the actual computation of the truth-value. So, rather than using logic as a meta-language describing facts, we are interested now in a particular internal realization of logic, be it in a computer or in a human. The fundamental difference is that truth values are not immediately given to us just because they apply by logical force to the terms but we have to calculate in each case which term has which truth value. This makes no difference with respect to classical logic. But the fact that computations may fail and that this failure itself is counted as a truth-value intertwines logic with its implementation. To take a concrete example consider the following part of a Pascal program.

(8)   if $x < 0$ then $1/(1-x) > 1+x$;

The second clause aborts if $x = 1$. However, the computer will never notice this, since the consequent is only considered if the antecedent is true; and the condition in the antecedent preempts this failure. In total, this sentence has no presupposition as far as the computer is concerned because it will under no circumstances fail. This, however, is due to two reasons. (a) The computer considers the consequent after the antecedent. (b) The computer drops the computation of the consequent in case the antecedent is not true. We could – in priciple – think of another strategy by which the consequent is checked first and the computation of the antecedent is dropped if the consequent is true. Then (8) will fail just in case $x = 1$. On the other hand, if the computer computes antecedent-to-consequent but looks at the consequent regardless of the antecedent, still it will fail if $x = 1$. So, both (a) and (b) are necessary.

We have isolated two properties of the standard computer implementations that produce the presuppositional behaviour of computers. One is the *directionality of computation* and the second is the *principle of economic computation*. If the computer implements no economy strategy the resulting logic is the so-called **Weak Kleene Logic** or **Bochvar's Logic**. It is characterized by the fact that any failure during a computation wherever it may arise will let the overall computation fail. The same logic will be derived even with the economy principle but with a different control structure. We can isolate four control structures; the first two are the uni-directional control structures *left-to-right* and *right-to-left*. The second are the bidirectional control structures; here, both directions are tried; the difference is whether the computation succeeds if only one branch succeeds (strong) or if both succeed (weak). These four cases correspond to four diacritics; $\overset{\shortmid}{\to}$ for left-to-right, $\overset{\triangleleft}{\to}$ for right-to-left, $\overset{\shortmid}{\to}$ for bidirectional and weak and $\overset{\diamond}{\to}$ for bidirectional and strong. The reader may check that this gives the following truth tables.

$\overset{\shortmid}{\to}$	T	F	U
T	T	F	U
F	T	T	U
U	U	U	U

$\overset{\triangleright}{\to}$	T	F	U
T	T	F	U
F	T	T	T
U	U	U	U

$\overset{\triangleleft}{\to}$	T	F	U
T	T	F	U
F	T	T	U
U	T	U	U

$\overset{\diamond}{\to}$	T	F	U
T	T	F	U
F	T	T	T
U	T	U	U

If we define the assignment relation $\leq$ between truth values in the obvious way ($\mathsf{U} \leq \mathsf{F},\mathsf{T}$) then the bidirectional connectives are defined from the unidirectional ones in the following way.

$$P \overset{!}{\to} Q = \min_{\leq}\{P \overset{\triangleright}{\to} Q, P \overset{\triangleleft}{\to} Q\}$$

$$P \overset{\diamond}{\to} Q = \max_{\leq}\{P \overset{\triangleright}{\to} Q, P \overset{\triangleleft}{\to} Q\}$$

There is also an interpretation that does not assume that computations may fail, i. e. that the basic predicates are partial, but nevertheless introduces three valued logic because it admits the possibility of broken channels in information transmission. Here we assume the connectives to be machines in a network which are supposed to answer queries. Once activated, the machines work the query backwards to the input channels. A single query can start an avalanche of queries which terminates in the variables. The latter we also regard as machines, working on no input; they are able to respond directly to a query. They can in principle give two answers, namely T and F. In that case, everything works as in classical logic.

Now suppose that the machines can also fail to respond because they are broken, because the connection is interrupted or because some machine fails to answer in due time. Our automaton $\vee$ somewhere in the network is thus faced with several options when the answers to the queries may turn out to be incomplete or missing at a time point. (a) It can wait until it receives the proper input; (b) It can use some higher order reasoning to continue in spite of an incomplete answer. (b) is the option with inbuilt economy principles and (a) is the option without. The (b) option branches into several distinct options. They are brought together as follows. We understand that in a binary function $f(\_1, \_2)$ there is an **information lock** between the first slot and the second slot. This lock has four positions. It can be closed (I), completely open ($\diamond$) and half-open, either to the right ($\triangleright$) or to the left ($\triangleleft$). These four positions determine in which way information about a received input i. e. about the value of the argument that is plugged in may flow. In closed position the left hand does not know what the right hand is doing. Even though P is received as T it is not known to the automaton when working at Q that P is true and it can therefore not know that it may now forget about the value of Q. It is still waiting. The same occurs if the information lock is open from right to left. The truth-values coincide with those given above. Of course we must now be careful with the boolean laws since it is not guaranteed that they hold. But the typical interdefinability laws of boolean logic hold for the connectives with similar control structure. We show some of them in the next lemma.

LEMMA 6.1   The following interdefinability laws hold:

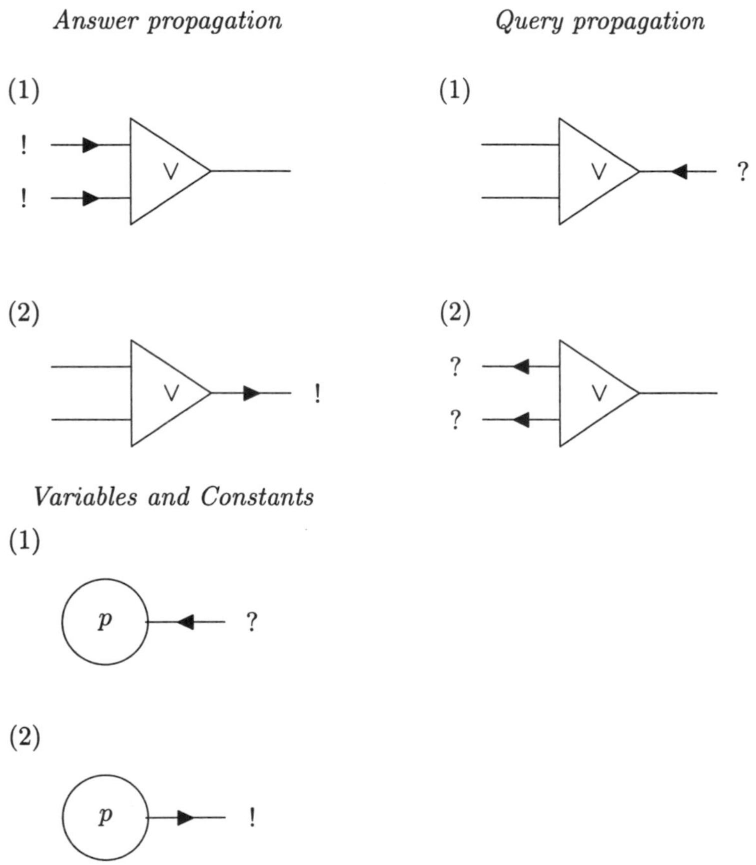

Figure 6.1
These pictures show the behaviour of logical automata. A connective may either compute an output answer (2) from the input answers (1) or an input query (2) from an output query. Variables transform output queries into output answers.

$$P \overset{\shortmid}{\vee} Q = \neg(\neg P \overset{\shortmid}{\wedge} \neg Q) \qquad P \overset{\shortmid}{\to} Q = \neg P \overset{\shortmid}{\vee} Q$$
$$P \overset{\triangleright}{\vee} Q = \neg(\neg P \overset{\triangleright}{\wedge} \neg Q) \qquad P \overset{\triangleright}{\to} Q = \neg P \overset{\triangleright}{\vee} Q$$
$$P \overset{\triangleleft}{\vee} Q = \neg(\neg P \overset{\triangleleft}{\wedge} \neg Q) \qquad P \overset{\triangleleft}{\to} Q = \neg P \overset{\triangleleft}{\vee} Q$$
$$P \overset{\diamond}{\vee} Q = \neg(\neg P \overset{\diamond}{\wedge} \neg Q) \qquad P \overset{\diamond}{\to} Q = \neg P \overset{\diamond}{\vee} Q. \qquad \blacksquare$$

It is instructive to see why this interpretation is sound even when the variables are assumed to be classical. The answer is not straightforward but simple. It is best understood with an example. We take $p \overset{\diamond}{\to} p$.

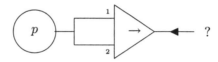

When a query is received, our automaton sends a query with both input channels. Suppose it receives answers as follows.

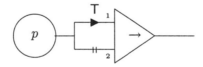

If we assume that the connection from 'p' to the second (= lower) input channel is broken, only the first channel receives the answer given by p. Yet the automaton cannot act. The reason is that the automaton does not know that the two input channels give the same answer; it only sees that they go out into the network but has no idea that they are systematically connected. So it is forced to wait for the second input. $p \overset{\diamond}{\to} p$ is thus not always true. The situation is comparable to that of a scientist concerned with the truth of two propositions r and s. Suppose that he does not know that they are in fact the same such as *Hesperus rises in the morning* and *Phosphorus rises in the morning*. So he cannot conclude that $r \overset{\diamond}{\to} s$ is true; in fact, only by long term experiment and/or statistical reasoning he may come to the conclusion that the two must be the same because the one is true exactly when the other is. Indeed, his local knowledge of the world compares directly to the limited knowledge of the $\overset{\triangleright}{\to}$-automaton.

Returning now to the issue of presupposition we notice that the explicit control structures give us the desired systematization of the logics; choosing uniformly the weak bidirectional connectives realizes Bochvar's Logic or Weak Kleene Logic, choosing the strong bidirectional connectives we get the Strong Kleene Logic and choosing the left-to-right interpretation gives us the typical computer implementations. In natural language the facts are not that simple. On closer inspection the connectives only show a certain tendency to be asymmetrical and left-to-right. But this default choice can be overridden as the bathroom sentences show.

(9)   Either the bathroom is upstairs or there is no bathroom.
(10)   If the bathroom is not upstairs there is none in the house.

It turns out that *or* tends to be weak and is directional only if there is a chance of cancelling a presupposition that would otherwise be inherited. Furthermore, *and* can be weak. Only *if ... then* shows a rather strong left-to-right tendency. But these are only rules of thumb.

6.4   A Formal Approach to the Projection Problem

In three-valued logic we agreed to let P be a proposition if $\beta(P) \in \{\mathsf{T}, \mathsf{F}\}$. So we have $D = \{\mathsf{T}\}$ and $U = \{\mathsf{U}\}$. An immediate consequence is that $P \triangleright Q$ iff $P \vee \neg P \vdash_3 Q$. For $P \vee \neg P$ is true iff P is either true or false, regardless of which of the four instantiations we choose. Thus if $P \triangleright Q$ then if $P \vee \neg P$ is true, P is either true or false and thus Q is true as well. And conversely. The same holds for classical logic. But since in classical logic no term can fail to be a proposition under no matter what valuation, the notion of presupposition becomes trivial.

PROPOSITION 6.1   In classical logic, P presupposes Q iff Q is a tautology.

Proof By definition, $P \triangleright_2 Q$ if Q is true whenever P is a proposition. Thus Q is always true, hence a tautology.   ∎
Only when we admit a third value the notion of presupposition starts to make sense. In the sequel we will indeed study presupposition in the context of three valued logic. We write $\triangleright$ for $\triangleright_3$.

PROPOSITION 6.2
(0) $P \triangleright Q$ iff $P \vee \neg P \vdash_3 Q$. ($\vee \in \{\overset{\shortmid}{\vee}, \overset{\triangleright}{\vee}, \overset{\triangleleft}{\vee}, \overset{\circ}{\vee}\}$.)
(i) If $P \triangleright Q$ and $P \equiv_3 P', Q \vdash_3 Q'$ then $P' \triangleright Q'$.
(ii) If $P \triangleright Q$ and $Q \triangleright R$ then $P \triangleright R$.
(iii) $P \triangleright Q$ iff $\neg P \triangleright Q$.
(iv) If $P \triangleright Q$ and $Q \triangleright P$ then $P \equiv_3 Q$ and P (as well as Q) is not falsifiable.

Proof (0) $\beta(P) \in \{T, F\}$ iff $\beta(P \vee \neg P) = T$. Hence if $P \vee \neg P$ is true, P is true or false and by $P \triangleright Q$, Q is true. Also if $P \vee \neg P \vdash_3 Q$ and P is true or false, $P \vee \neg P$ is true and so Q is true. (i) Let $P \triangleright Q$, $P' \equiv_3 P$ and $Q \vdash_3 Q'$. Assume $\beta(P') \in \{T, F\}$. Then $\beta(P) \in \{T, F\}$ as well. Thus $\beta(Q) = T$; now by $Q \vdash_3 Q'$ also $\beta(Q') = T$. (ii) If $Q \triangleright R$ then a fortiori $Q \vdash_3 R$ and by (i) $P \triangleright R$. (iii) $P \triangleright Q$ iff $P \overset{\diamond}{\vee} \neg P \vdash_3 Q$ iff $\neg\neg P \overset{\diamond}{\vee} \neg P \vdash_3 Q$ iff $\neg P \triangleright Q$. (iv) Clearly, $P \triangleright Q$ implies $P \vdash_3 Q$ and $\neg P \vdash_3 Q$ and $Q \triangleright P$ implies $Q \vdash_3 P, \neg Q \vdash_3 P$. Thus $P \vdash_3 Q \vdash_3 P$. Furthermore, suppose that P is false; then Q is true and so P must be true as well. Contradiction. Similarly, Q cannot be false. Finally, if $\beta(P) = U$, then Q cannot be true, otherwise P is true. Q is not false either, hence $\beta(Q) = U$ as well. Dually, if $\beta(Q) = U$ then also $\beta(P) = U$. This shows $P \equiv_3 Q$. ∎

Hence, if we consider the set $\mathfrak{N}$ of all nonfalsifiable formulae then $\triangleright$ turns out to be irreflexive and transitive on $\mathfrak{N}$. Let us now analyse the formal behaviour of presuppositions in languages of three valued logic. To this end we define a binary connective $\downarrow$ by

$\downarrow$	T	F	U
T	T	F	U
F	U	U	U
U	U	U	U

PROPOSITION 6.3 $(Q \downarrow P) \triangleright Q$. Moreover, $P \triangleright Q$ iff $Q \downarrow P \equiv_3 P$.

Proof The first claim is easy to verify. We turn to the second. ($\Rightarrow$) If $Q \downarrow P$ is true then P is true. If P is true then by $P \triangleright Q$ also Q is true, hence $Q \downarrow P$ is true. If $Q \downarrow P$ is false, P is false. If P is false, Q is true by $P \triangleright Q$ and so $Q \downarrow P$ is false. ($\Leftarrow$) If P is true, $Q \downarrow P$ is true, so Q is true. This shows $P \vdash_3 Q$. If P is false then $Q \downarrow P$ is false as well; and so Q is true showing $\neg P \vdash_3 Q$. So, $P \triangleright Q$. ∎

It is possible to define a unary connective ∇ by $\nabla P := P \downarrow P$. It has the following truth-table.

	∇
T	T
F	U
U	U

The idea of such a connective is due to D. Beaver. This unary connective allows to define presuppositions in a similar way as $\downarrow$. This due to the fact established in the next theorem.

PROPOSITION 6.4 $Q \downarrow P \equiv_3 \nabla Q \overset{\downarrow}{\wedge} P \equiv_3 \nabla Q \overset{\triangleright}{\wedge} P$.

Proof The formulae are truth-equivalent. For all three are true iff both P and Q are true. $Q \downarrow P$ is false iff P is false and Q is true iff ∇Q is true and P is false iff $\nabla Q \overset{\shortmid}{\wedge} P$ is false iff $\nabla Q \overset{\triangleright}{\wedge} P$ is false.   ∎

To be precise: as soon as $\overset{\shortmid}{\wedge}$ or $\overset{\triangleright}{\wedge}$ or $\overset{\triangleleft}{\wedge}$ are definable, $\downarrow$ is definable from ∇ and one of the three. It is therefore a matter of convenience whether we use $Q \downarrow P$ or some definition using ∇.

LEMMA 6.2   Let $*, \circ \in \{\overset{\shortmid}{\wedge}, \overset{\triangleright}{\wedge}, \overset{\triangleleft}{\wedge}, \overset{\circ}{\wedge}\}$. Then $\nabla P * \nabla Q \equiv_3 \nabla (P \circ Q)$.

Proof $\nabla P * \nabla Q$ can never be false since ∇P, ∇Q can never be false. Hence, we need to check only equivalence in truth. But $\nabla P * \nabla Q$ is true iff ∇P, ∇Q are true iff P and Q are true iff $P \circ Q$ is true iff $\nabla (P \circ Q)$ is true.   ∎

LEMMA 6.3 $R \downarrow (Q \downarrow P) \equiv_3 (R \downarrow Q) \downarrow P$. Furthermore, if $* \in \{\overset{\shortmid}{\wedge}, \overset{\triangleright}{\wedge}, \overset{\triangleleft}{\wedge}, \overset{\circ}{\wedge}\}$ then $R \downarrow (Q \downarrow P) \equiv_3 (Q * R) \downarrow P$.

Proof We show the second claim first. $R \downarrow (Q \downarrow P) \equiv_3 \nabla R \overset{\shortmid}{\wedge} (Q \downarrow P) \equiv_3 \nabla R \overset{\shortmid}{\wedge} \nabla Q \overset{\shortmid}{\wedge} P \equiv_3 \nabla (Q * R) \overset{\shortmid}{\wedge} P \equiv_3 (R * Q) \downarrow P$. Now for the first. $R \downarrow (Q \downarrow P) \equiv_3 \nabla R \overset{\shortmid}{\wedge} (Q \downarrow P) \equiv_3 \nabla R \overset{\shortmid}{\wedge} (\nabla Q \overset{\shortmid}{\wedge} P) \equiv_3 \nabla (\nabla R \overset{\shortmid}{\wedge} Q) \overset{\shortmid}{\wedge} P \equiv_3 (R \downarrow Q) \downarrow P$.   ∎

The main theorem of this paragraph deals with the projection problem. Our solution is completely formal and bears only on the logical properties of presuppositions. We will approach the projection problem via *normal forms*.

DEFINITION 6.2   Let $\mathbb{L}$ be a language of boolean connectives with locks and $\mathbb{L}^{\downarrow}$ its expansion by $\downarrow$. We say a formula $P \in \mathbb{L}^{\downarrow}$ is in **presuppositional normal form** (**pnf**) if P is syntactically equal to $Q_2 \downarrow Q_1$ for some $Q_1, Q_2 \in \mathbb{L}$.

THEOREM 6.1   For each $P \in \mathbb{B}_3$ there exists a $\pi(P) \equiv_3 P$ which is in presuppositional normal form.

Proof The following are valid statements:

$(as \downarrow)$ $(R \downarrow Q) \downarrow P \quad \equiv_3 (R \wedge Q) \downarrow P$

$(di \downarrow)$ $R \downarrow (Q \downarrow P) \quad \equiv_3 (R \wedge Q) \downarrow P$

$(ne \downarrow)$ $\neg (Q \downarrow P) \quad \equiv_3 Q \downarrow (\neg P)$

$(co \downarrow)$ $(Q \downarrow P) \overset{\shortmid}{\wedge} R \quad \equiv_3 Q \downarrow (P \overset{\shortmid}{\wedge} R)$

$(Q \downarrow P) \overset{\triangleright}{\wedge} R \quad \equiv_3 Q \downarrow (P \overset{\triangleright}{\wedge} R)$

$(Q \downarrow P) \overset{\triangleleft}{\wedge} R \quad \equiv_3 (R \overset{\triangleright}{\to} Q) \downarrow (P \overset{\triangleleft}{\wedge} R)$

$(Q \downarrow P) \overset{\circ}{\wedge} R \quad \equiv_3 (R \overset{\circ}{\to} Q) \downarrow (P \overset{\circ}{\wedge} R)$

These statements if read from left to right, provide an algorithm for deriving a pnf for a formula containing negation and conjunctions. With the fact that all boolean connectives with locks can be represented with just negation and the conjunctions with locks, we can effectively reduce all terms by first eliminating the other booleans and then reducing according to the equivalences shown above. The task is thus to prove them. Lemma 6.3 showed $(as \downarrow)$ and $(di \downarrow)$, $(ne \downarrow)$ is not difficult. For the laws $(co \downarrow)$ observe the fact that all left hand sides and all right hand sides are true iff P, Q and R are true. So it is enough to establish that they receive U (or F) under the same valuations. **(1)** $(Q \downarrow P) \overset{|}{\wedge} R$ is U iff $Q \downarrow P$ is U or R is U iff either Q is not T or P is U or R is U iff Q is not T or $P \overset{|}{\wedge} R$ is U iff $Q \downarrow (P \overset{|}{\wedge} R)$ is U. **(2)** $(Q \downarrow P) \overset{\triangleright}{\wedge} R$ is U iff either $Q \downarrow P$ is U or $Q \downarrow P$ is T and R is U iff either Q is not T or P is U or P, Q are T and R is U iff Q is not T or $P \overset{\triangleright}{\wedge} R$ is U iff $Q \downarrow (P \overset{\triangleright}{\wedge} R)$ is U. **(3)** $(Q \downarrow P) \overset{\triangleleft}{\wedge} R$ is U iff R is U or R is T and $Q \downarrow P$ is U iff R is U or R is T and: either P is U or Q is not T iff either R is U or R is T and P is U or R is T and Q is not T; $(R \overset{\triangleright}{\rightarrow} Q) \downarrow (P \overset{\triangleleft}{\wedge} R)$ is U iff $R \overset{\triangleright}{\rightarrow} Q$ is not T or $P \overset{\triangleleft}{\wedge} R$ is U iff either R is U or R is T but Q is not T or R is U or R is T and P is U iff either R is U or R is T and P is U or R is T and Q is not T. **(4)** $(Q \downarrow P) \overset{\diamond}{\wedge} R$ is F iff either $Q \downarrow P$ is F or R is F iff either Q is T and P is F or R is F; $(R \overset{\diamond}{\rightarrow} Q) \downarrow (P \overset{\diamond}{\wedge} R)$ is F iff $R \overset{\diamond}{\rightarrow} Q$ is T and $P \overset{\diamond}{\wedge} R$ is F iff P is F and Q is T or P is F and R is F or R is F iff R is F or Q is T and P is F. $\blacksquare$
Now that projection of presuppositions is formally defined let us see how it helps in finding out the presuppositions of an arbitrary formula. Let us define the **generic presupposition** of P to be such a Q that $P \triangleright R$ iff $Q \vdash_3 R$. Such a generic presupposition always exists; just take $Q = P \overset{|}{\vee} \neg P$; moreover, it is unique up to deductive equivalence. Because if $\widehat{Q}$ is another such generic presupposition then $Q \vdash_3 \widehat{Q}$ as well as $\widehat{Q} \vdash_3 Q$. It does not follow, however, that generic presuppositions are equivalent! By Proposition 6.2, $P \overset{|}{\vee} \neg P, P \overset{\triangleleft}{\vee} \neg P, P \overset{\triangleright}{\vee} \neg P$ and $P \overset{\diamond}{\vee} \neg P$ are all generic presuppositions. These easy solutions have a serious disadvantage. If P contains $\downarrow$, so does $P \overset{|}{\vee} \neg P$, but we like to have a generic presupposition free of $\downarrow$.

THEOREM 6.2 Let $Q_2 \downarrow Q_1$ be a presuppositional normal of P. Then $(Q_1 \overset{|}{\vee} \neg Q_1) \overset{|}{\wedge} Q_2$ is a generic presupposition of P and free of $\downarrow$.

Proof It is enough to show $P \overset{|}{\vee} \neg P \vdash_3 (Q_1 \overset{|}{\vee} \neg Q_1) \overset{|}{\wedge} Q_2 \vdash_3 P \overset{|}{\vee} \neg P$.

$$P \stackrel{\shortmid}{\vee} \neg P \quad \equiv_3 \quad Q_2 \downarrow Q_1 \stackrel{\shortmid}{\vee} \neg(Q_2 \downarrow Q_1)$$
$$\equiv_3 \quad Q_2 \downarrow Q_1 \stackrel{\shortmid}{\vee} Q_2 \downarrow(\neg Q_1)$$
$$\equiv_3 \quad (\nabla Q_2 \stackrel{\shortmid}{\wedge} Q_1) \stackrel{\shortmid}{\vee} (\nabla Q_2 \stackrel{\shortmid}{\wedge} \neg Q_1)$$
$$\equiv_3 \quad \nabla Q_2 \stackrel{\shortmid}{\wedge} (Q_1 \stackrel{\shortmid}{\vee} \neg Q_1)$$

Together with $\nabla Q_2 \vdash_3 Q_2 \vdash_3 \nabla Q_2$ the claim quickly follows.   ∎

Now let $\mathbb{L}$ be a language of boolean connectives with locks, and as above $\mathbb{L}^{\downarrow}$ its $\downarrow$-extension. If $P \in \mathbb{L}^{\downarrow}$, denote by $P[Q \downarrow \overline{p}/\overline{p}]$ the result of uniformly substituting $Q \downarrow p$ for p for every variable p of P.

THEOREM 6.3 $Q \downarrow P \equiv_3 P[Q \downarrow \overline{p}/\overline{p}]$.

Proof $Q \downarrow P$ is true iff P and Q are true iff $P[Q \downarrow \overline{p}/\overline{p}]$ is true. For if Q is not true, all $Q \downarrow p$ are undefined and so is $P[Q \downarrow p/\overline{p}]$. But by induction, P is undefined if all variables are U. Hence Q must be true and then $P[Q \downarrow \overline{p}/\overline{p}]$ reduces to P. Thus $Q \downarrow P$ is false iff P is false and Q is true iff $P[Q \downarrow \overline{p}/\overline{p}]$ is false.   ∎

6.5   Separation: Internal or External?

At first glance, separation does not look problematic; but a short investigation into the formal prerequisites that make separation possible will produce surprises. To begin we investigate the relationship between two- and three-valued logic. Since in our situation we have defined our three-valued logics as extensions of classical logic it seems straightforward to imitate two-valued logic within three valued logic but more problematic to interpret three-valued logic in two valued logic. But so it only seems.

To begin let us assume that we have some variable p. A priori, p can assume three values, T, F and U. Now define operations $p^{\triangle}$ and p^{∇} with the following properties. (i) Both $p^{\triangle}$ and p^{∇} are propositions, i.e., have only classical truth values. (ii) $p \equiv_3 p^{\nabla} \downarrow p^{\triangle}$. $p^{\triangle}$ and p^{∇} by definition solve the separation problem. It is easy to show, however, that no system using connectives with whatever control structure can produce p^{∇} and $p^{\triangle}$. The reason is simply that any term of that system assumes U if all variables are U. There is no way to define propositions from sentences that are not necessarily propositions themselves. Nevertheless, we might argue that we have considered a system that is too weak; we might, for example, add weak negation.

	$\sim$
T	F
F	T
U	T

Then $p^\triangledown \equiv_3 \sim(p \overset{\shortmid}{\vee} \neg p)$ and $p^\triangle \equiv_3 \sim\sim p$. In that case, these two functions become definable. We can on the other hand define weak negation from the assertion function by $\sim p \equiv_3 \neg p^\triangle$. Furthermore, $\sim p \equiv_3 p^\triangledown \overset{\triangleright}{\vee} \neg p$ and so in principle one of the three functions is sufficient to define the others (given enough of the other connectives).

The above arguments show that it is not clear that we can separate any sentence *language internally*. It is, however, always possible to separate sentences *language externally* by stipulating two functions $(-)^\triangledown$ and $(-)^\triangle$ that produce the assertion and proposition of that sentence in another language. If we add certain functions (e.g., weak negations) to our language the external functions can be mimicked internally as we have seen, but the system with external functions has some conceptual advantages for natural language analysis. With the three valued projection algorithm we can formulate principles of $^\triangledown$- and $^\triangle$- percolation. For example,

$$
\begin{aligned}
(P \overset{\shortmid}{\wedge} Q)^\triangledown &= P^\triangledown \wedge Q^\triangledown \\
(P \overset{\triangleright}{\wedge} Q)^\triangledown &= P^\triangledown \wedge (P^\triangle \to Q^\triangledown) \\
(P \overset{\triangleleft}{\wedge} Q)^\triangledown &= Q^\triangledown \wedge (Q^\triangle \to P^\triangledown) \\
(P \overset{\diamond}{\wedge} Q)^\triangledown &= [P^\triangledown \wedge (P^\triangle \to Q^\triangledown)] \vee [Q^\triangledown \wedge (Q^\triangle \to P^\triangledown)]
\end{aligned}
$$

$$
\begin{aligned}
(P \overset{\shortmid}{\wedge} Q)^\triangle &= P^\triangle \wedge Q^\triangle \\
(P \overset{\triangleright}{\wedge} Q)^\triangle &= P^\triangle \wedge Q^\triangle \\
(P \overset{\triangleleft}{\wedge} Q)^\triangle &= P^\triangle \wedge Q^\triangle \\
(P \overset{\diamond}{\wedge} Q)^\triangle &= P^\triangle \wedge Q^\triangle
\end{aligned}
$$

Notice that the lower equations do not contain the three valued-connectives; this is because we use an external separation in two-valued logic. If we separate internally we can use any of the control equivalents on the right hand side since only the classical values count. Notice also that the assertive part is rather regular in its behaviour.

The discussion on separation becomes less academic when we look at semantics. With few exceptions, semantical theories use two valued logic: Montague Semantics, Discourse Representation Theory, Boolean Semantics, etc. [14] presents an account of presupposition within DRT. Interestingly, however, he makes no attempt to solve the problem of how to mediate between two- and three-valued logic. Of course, he doesn't see such a problem arise because he views presuppositions as kinds of anaphors so that there is no projection, just allocation. But this means that he closes his eyes in front of some problems. Firstly, there is a marked difference between presuppositions being allocated *as presuppositions* and presuppositions being allocated as *antecedents* or *conjuncts*; we will return to this later. Secondly, the additional conceptual layer of two-valued predicates needs arguing for; it is comfortable to have it but arguably quite unnecessary. Let us consider an example.

(9)   Hilary is unmarried.
($9^\dagger$) **_unmarried′_**(h)
($9^\ddagger$) **unmarried′**(h)

A direct translation of (9) is ($9^\dagger$) which uses a three-valued prediate $\lambda x.$**_unmarried′_**(x) which we simply write as **_unmarried_**. The two-valued equivalent is $\lambda x.$**unmarried′**(x) or simply **unmarried**. The two are not the same; it is clear that we would not say of a desk, a star or an ant that it is unmarried. Neither would we say this of a six-year old child. In logic we say that *unmarried* is type restricted. It applies only to objects of a certain type. This type restriction is a presupposition since it stable under negation. In the same way *bachelor* is type restricted to all objects that are male and satisfy the type restriction of *unmarried*. There is a whole hierarchy of types which are reflected in the net of elementary presuppositions carried by lexical items. The predicate **unmarried** is not type restricted hence only truth-equivalent to **_unmarried_**. Indeed, we have exactly **unmarried** $\equiv_3$ **_unmarried_**$^\Delta$. It is difficult to verbalize this. To say that (9) asserts ($9^\ddagger$) is no proof that such an entity exists; language does not allow to introduce a simple way to express ($9^\ddagger$) without introducing standard presuppositions. This claim on my side needs arguing, of course. Prima facie nothing excludes there being such a predicate conforming to **unmarried**. I would, however, not go as far as that. All I am saying is that there is no straightforward, logical procedure to verbalize the assertional and presuppositional part of a sentence. The operations $(-)^\Delta$ and $(-)^\nabla$ are merely theoretical devices, and weak negation does not exist contrary to what is sometimes claimed. There is no way of saying simply (9) or (10) to mean ($10^\ddagger$). An argument using (11) as evidence for weak negation begs the question because it is clear that the added conflicting material serves to identify the presupposition that is cancelled. If it is dropped the negation is interpreted as strong. In addition, I still find such examples of questionable acceptability.

(10) Hilary is not married.
($10^\ddagger$) $\neg$**married′**(h)
(11) ? Hilary is not married; she is only six years old!

In a similar we can see that spelling out the two-valued presupposition of *unmarried* is not straightforward. In a first attempt we write **_unmarried_**$^\nabla$ = **_human_** $\wedge$ **_adult_** but we find that *adult* itself has type restrictions.

6.6   Dependencies and Allocation

[14] recently offered a rather detailed account of a theory of allocation. We will not try
to improve here on the empirical coverage of that theory; rather we will note some defi-
ciencies of the theory itself. Firstly, this theory uses DRT and classical logic. This might
be a surprising diagnosis because he explicitly states that the Frege-Strawsonian theory
of presupposition can be reinstalled by using a truth-value gap. Yet, he claims that this
gap arises from violating explicit binding constraints and so is in effect reducible to binding. Furthermore, van der Sandt makes use of separation. We have argued against that
earlier; I will take the opportunity to discuss a further disadvantage of internal separation. If presuppositions and assertions of words such as *bachelor* are separable internally
and thereby in principle arbitrarily assignable, why does this extra freedom never get
exploited? It is namely possible to define a different concept, say, *lachelor*, which has
the truth conditions of a bachelor but is presupposes that the object is a living thing.
Thus $lachelor \equiv living \downarrow bachelor$; *lachelor* is truth-equivalent with **bachelor**
but evidently not falsity equivalent. Yet such a concept is not lexicalized in the language
and it is quite difficult to imagine such a concept. So it seems that presuppositions are
just part of the concept and not dissociable from it not the least because we have argued
against an additional two-valued interpretation language for reasons of economy. Hence,
the story of reallocation (which I also used in the introduction) is just a bad metaphor.
The presupposition cannot be freed from the concept; this seems to be at least intuitively
accepted (see [6]) even though this and other logical implementations do not take notice
of that fact. Hence, if presuppositions do not move, they have in some sense to be *copied*.
We can understand this as follows. Knowing that at a certain point we have to satisfy
certain presuppositions (which are directly given by the concept) we decide to accommodate this presupposition at some place so that the presupposition in question is satisfied
at the point we come to evaluate it. It is, however, important (and has been overlooked
by van der Sandt) that the reallocated presupposition functions as a presupposition. In
the semantics we thus need to postulate the connectives ∇ or $\downarrow$ as primitives in order
to guarantee that the inserted material can function as a presupposition not just as an
assertion.

The next problem to be considered is the reduction to binding conditions. Probably
this can be made meaningful in the following way.

```
        program NULL;          program NULL2;
          begin                  N : integer;
            N := 0;                begin
          end.                       N := 0;
                                   end.
```

If we compare these two Pascal programs we see that the left program will fail because the variable N has not been declared. In the other this has been done and so it operates successfully. In the same way we can understand the functioning of the upper part of the box in DRT. It makes a DRS to undefined if a variabel is used but not defined in the head section. So in the DRSs below the left is ill-formed while the one to the right is well-formed.

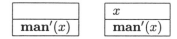

This would be enough to exclude unbound occurrences if we assume some extra control priciples handling composite structures; however, van der Sandt chose to translate lexical items in such a way that the binding conditions are satisfied and artificially adds the clause $\alpha_0 =?$ to make binding required. I do not see how this can be superior to an account where the variable mechanism itself is exploited. Prior to inserting a DRS for a lexical item we have to choose anyway the variables we are going to insert and we can rely totally on this mechanism to create the preusppositional effect.

Thirdly, it can be demonstrated that presuppositions are as matter of fact not reducible to anaphor resolution. There are obvious arguments against this. One is that there are presuppositions that simply have no variable to be bound. These are easy cases. A trickier example is this one.

(12)   If John buys a car he removes the spare wheel.
(13)   ? If John buys a car he removes his spare wheel.

(12) could in principle be interpreted in two ways; the presupposition can stay or be reallocated at the top-level. People prefer the first alternative simply because cars tend to come with spare wheels. On the other hand it is possible that in certain contexts (12) presupposes that there is a definite spare wheel, e.g., when the prvious discourse established a particular spare wheel already. So the choice where the presupposition is accommodated is not governed by binding facts but by some kind of pragmatic *dependency*. Choosing spare wheels to be dependent in some form on cars we opt for a local reading of the presupposition. By default, however, presuppositions or objects are independent. Specific knowledge, in this case about cars and spare wheels, is required to establish such a dependency. We can use the context to create such dependencies or overrun them. Sentence (13), however, is the interesting bit if the argument. The theory of [14] has no means to tell us why this sentence is so odd. It is not the choice of the locus of the presupposition that is at stake but the question whether the spare wheel is actually identifiable with the one that arises naturally with the car if the presupposition is reallocated at the top. This has nothing to do with binding; we can in principle have it either way.

A theory of allocation must be pragmatic in nature as we have seen. Van der Sandt agrees with that because he limits the locus not only by binding conditions but also by pragmatic factors. But with respect to the last he remains rather vague. He only considers cases in which there appears a logical conflict if the locus of the presupposition is too high. This leaves the impression that logic plays a significant role as a determining factor. I am tempted to say that it does not. It seems to me that this theory should be based on a theory of *dependencies*. I used this term earlier but let me be a bit more precise here what I mean by that. Dependencies are relations that hold between objects, between concepts or between objects and concepts etc. In logic, when we use arbitrary objects (see [3]) an object that is freed from an existential quantifier depends on all the free variables or constants of the formula. So, $(\forall x)(\exists y)\varphi(x,y)$ does not imply $(\exists y)(\forall x)(\varphi(x,y))$ because if we were to free objects from the quantifiers we get dependencies that block the reintroduction of the quantifier for an object. This goes as follows. We first assume an a such that $(\exists y)\varphi(a,y)$. Then we take a b such that $\varphi(a,b)$. The conclusion that therefore $(\forall x)\varphi(x,b)$ is not true because b depends on a. In this way the theory by Fine shows how reasoning with quantifiers is reducible with objects. The dependencies are the main thing that we have to memorize if we want to reason correctly. Notice that if we manage to prove that b actually does not depend on the choice of a the above reasoning could be carried on and the implication $(\forall x)(\exists y)\varphi(x,y) \rightarrow (\exists y)(\forall x)\varphi(x,y)$ holds. The observable is therefore the dependency of b on a. How we recognize this dependency is quite another story but it certainly limits the reasoning. In totally the same way we understand the mechanism of reallocation. *Dependency* is a primitive notion; by some means we come to recognize that the truth of some proposition or the referent of some description is dependent on some other. It is then a consequence that the dependent description or proposition cannot be processed before the one on which it depends and this in turn explains why the locus of a presupposition must be inside the scope of all things on which the presupposition depends. A final illustration is this sentence.

(14) As long as the moon wanders around the earth fishermen will love the tide.

It is known (not to all of us) that the moon creates the tide. So, *the tide* ceases to refer as soon as moon stops wandering. Hence the locus of the presupposition is its origin. But if we fail to know (or notice) that we will read this as saying that there is a tide whether or not the moon is wandering around the world. The two differ not only logically; in the latter reading we are led to think that it is the moon which induces the love of the fishermen for the tide in some way. the latter therefore leads us to see a dependency between the moon's wandering around the earth and fishermen's love of the tide whereas in the first case no dependency is seen between the moon's wandering and the existence of the tide.

6.7 Conclusion

In comparing possible extensions of two-valued logic to three-valued logic using explicit control structures we have managed to give an account of standard presuppositional behaviour of computers or mathematicians. We have boosted this up to natural language semantics by assuming possible reallocation of presupposition. The latter extension touches on pragmatics and is therefore not easily spelt out in detail. We will in this last section evaluate the pros and cons not of the theory of allocation but of the interpretation in three-valued logic.

It has been noted that standard three-valued logics for presupposition can be rephrased with the help of the rules of *context change*. Underlying that is the notion of presupposition as failure. This would lead to Bochvar's Logic was it not the case that the evaluation procedure is spelt out differently. Let us study the following two clauses (cf. [6]).

($\mathbf{lc} \rightarrow$) If the local context for *if A then B* is X, the local context for A is X and the local context for B is $X \cup \{A\}$.

($\mathbf{lc} \vee$) If the local context for *A or B* is X then the context for A is X and the local context for B is $X \cup \{\neg A\}$.

Each of A and B may carry their own presuppositions but in ($\mathbf{lc} \rightarrow$) the presuppositions for B are evaluated only in those situations where not only X holds but also A. Taking this together with the standard two-valued interpretation of $\rightarrow$ yields the truth-tables for $\overset{\triangleright}{\rightarrow}$. This is not hard to check. Similarly, ($\mathbf{lc} \vee$) leads to $\overset{\triangleright}{\vee}$. The problems that arose were that (1) the local context is not fixed by the connective and (2) one cannot freely assign any rule of local context to a connective. We cannot, for example, choose to take $X \cup \{\neg A\}$ as the context for B in ($\mathbf{lc} \rightarrow$). [6] is particularly worried by this. But the problem is that too much is specified in the rules of local context. We have seen earlier that the control structure is enough; so rather than anticipating the actual context against which B is evaluated we only say that A has to be evaluated first and B is evaluated against the context that results from X by adding the condition that must be satisfied if the computer is about to process B. This readily explains the difference between the rules of local context of ($\mathbf{lc} \rightarrow$) and ($\mathbf{lc} \vee$). We have seen that the control structures are independent of the connective and that the connective plus the control structure yield a definite truth-table. Each of the possible combinations is realized in language. As examples we study (15) and (16). If we consider all four possible options we see that (15) and (16) are free of presupposition if the context rules are spelled out as ($\mathbf{lc'} \vee$) and ($\mathbf{lc'} \rightarrow$). They sound rather circular but in fact reflect the control strategy of $\overset{\diamond}{\vee}$ and $\overset{\diamond}{\rightarrow}$.

(15) Either John has started smoking or he has just stopped smoking.

(16) If John hasn't started smoking he has just stopped smoking.

(lc'∨) If X is the local context for *either A or B* then the local context for A is $X \cup \{\neg B\}$ and the local context for B is $X \cup \{\neg A\}$.

(lc' →) If X is the local context for *if A then B* then the local context for A is $X \cup \{\neg B\}$ and the local context for B is $X \cup \{A\}$.

We can perform the same trick with implication and thereby force a symmetrical reading or the implication. It remains to be seen, however, what exactly determines this choice of the control structure. This we have not been able to establish nor the conditions under which it may take place at all; nor how this relates with the dependencies.

References

[1]   Blamey, Steven: *Partial Logic*, in: Gabbay & Guenthner (eds.): Handbook of Philosophical Logic, vol. 3, Reidel, Dordrecht, 1986, 1 - 70

[2]   Bochvar, D. A.: *On a Three-valued Logical Calculus and its Applications to the Analysis of the Paradoxes of the Classical Extended Functional Calculus*, translated by Merrie Bergman, History and Philosophy of Logic 2(1981), 87 - 112

[3]   Fine, Kit: *Natural Deduction and Arbitrary Objects*, Journal of Philosophical Logic 14(1985), 57 - 107

[4]   van Fraassen, B.: *Presupposition, Implication and Self-Reference*, Journal of Philosophy 65(1968), 135 - 152

[5]   Hayes, P.: *Three-valued Logic and Computer Science, Part I: Propositional Calculus and 3-valued inference*, Ms. of the University of Essex, 1975

[6]   Heim, I.: *Presupposition Projection*, Workshop on Presupposition, Lexical Meaning and Discourse Processes, 1990

[7]   Karttunen, L.: *Presupposition and linguistic context*, Theoretical Linguistics 1(1979), 181 - 194

[8]   Karttunen, L. and Peters, S.: *Conventional implicature*, in: Oh and Dinneen (eds.): *Synatx and Semantics 11: Presupposition*, Acacdemic Press, New York, 1979, 1 - 56

[9]   Kempson, R. M.: *Presupposition and the delimitation of semantics*, Cambridge University Press, Cambridge, 1975

[10]   Langendoen, D. H., Savin, L.: *The projection problem for presuppositions*, in: Fillmore & Langendoen: *Studies in Linguistic Semantics*, Holt, New York, 1971, 5.2 - 6.2

[11]   Langholm, Thore: *Partiality, Truth and Persistence*, CSLI Lecture Notes No. 15, 1989

[12]   Rautenberg, W.: *Klassische und nichtklassische Aussagenlogik*, Vieweg Verlag, Wiesbaden, 1979

[13]   van der Sandt, Rob A.: *Presupposition and Context*, Croom Helm Linguistic Series, London, 1988

[14]   van der Sandt, R. A.: *Anaphora and accommodation*, Workshop on Presupposition, Lexical Meaning and Discourse Processes, 1990

[15]   Urquhart, Alasdair: *Many-valued logic*, in: Gabbay & Guenthner (eds.): Handbook of Philosophical Logic, vol. 3, Reidel, Dordrecht, 1986

[16]   Visser, Albert: *Actions under Presuppositions*, Preprint No. 76, Logic Group of the Dept. of Philosophy, University of Utrecht, 1992

[17]   Wójcicki, A.: *Theory of logical calculi*, Kluwer, Dordrecht, 1988

7 Classification Domains and Information Links: A Brief Survey

Lawrence Moss and Jerry Seligman

7.1 Introduction

There are several issues which every theory of information and information-flow must address. Among these are the role of perspective and background on inference. There is an intuition that information is relative to the observer or recipient; but a thoroughgoing relativism on this point seems to contradict another intuition, that information is an inherent property of parts of the world and facts which they might support. A second issue is to clarify what it means to say that sources of information are usually reliable but sometimes fallible. Further, a theory should explain the relation of logical deduction to other types of information-flow, such as observation.

This chapter discusses work related to the construction of an account of information which is intended to address some of these points. We survey of some of the motivations, formal results, and applications of the study of classification domains and information links to the modeling of information-flow. Sections 7.2 to 7.6 are a digest of work in this direction by Jon Barwise and Jerry Seligman.

The central notion of "classification domain" is introduced in Section 7.2. Then, various ways of modeling the flow of information in a classification domain are discussed in Section 7.3. In Section 7.4, we settle on the definition of an "information link" between classification domains. An axiomatic characterization of the connection between information flow and logic is presented in Section 7.5 and this is related to an algebraic characterization of information links in Section 7.6.

The matter of situating deduction on the very broad field of information-flow, is not the only point of contact between logic and the flow of information, the keywords in the title of this volume. A theory of information-flow, as a mathematical enterprise, generates logics of its own. We illustrate this in Section 7.7 with a recent developments, the modal logic for topological reasoning of Moss and Parikh [8].

Another, rather different enterprise is to see entire fields of mathematics as *information-driven*. We have in mind logic, as well as parts of algebra and the basics of general topology. There is a feeling that the basics of these subjects can be understood as embodiments of intuitions about information at some level. (This is in contrast with, say, geometry and analysis, which can be considered as fields based on very different kinds of intuitions.) The point here is to see whether the rather loose feeling can be put on a firmer foundation once we have a theory of information in place. Part of the point of Section 7.7 is to solidify the intuition that point-set topology has something to do with information and observation.

7.2 Classification Domains

A **classification domain** (CD) is a triple $\langle S, T, \models \rangle$ consisting of two sets, S and T, together with a subset $\models$ of $S \times T$. We think of S as a collection of *sites*, T as a collection of *types*, and the relation $\models$ as *classification* of sites by types.

As the name suggests, classification domains are a formalization of the concept of classifying. Of course, the simple definition will only lead to interesting and useful structure when it is accompanied by other definitions that sharpen and refine it. The subject matter of this chapter, information links, is mainly the study of the *morphisms* of classification domains. Before we turn to it, we present a number of examples and discuss the main applications.

The first example is from traditional logic; it shows where the notation comes from. Here we begin with a *language* $\mathcal{L}$. Then S is taken to be some set of structures for $\mathcal{L}$, T is some set of $\mathcal{L}$-sentences, and the relation $\models$ is that given by the logic.

The second example is also mathematical. Let $\mathcal{T} = \langle X, \mathcal{O} \rangle$ be a topological space, a set X of *points* together with a family $\mathcal{O}$ of *opens*. The opens are subsets of X, and the entire collection $\mathcal{O}$ must satisfy some closure condition. We think this as a CD by taking the points as the sites, the opens as the types, and then the membership relation is the classification. The point of view that topological spaces may be understood as classifications is not always part of presentations of topology, but it is part of the working vocabulary of theoretical computer scientists who use topology in modeling information. The idea is that the opens are properties of the points. Moreover, the points in many cases correspond to something observable, and often one wants to reason about something ideal. (For example, when thinking about machines accepting input from the outside world, finite strings are all we observe, and infinite input streams are the ideal.) The completeness constructions of topology suggest how to construct models of the ideal elements from the space itself. This is a kind of pastiche of the uses of topology and domain theory in several areas. For more on it, cf., e.g., Vickers [12].

For a very different example, consider the weather. Let S be the set of spatio-temporal situations, and let T be the set of pairs (a, b) of rational numbers with $a < b$. Finally, let $s \models (a, b)$ iff in s, the temperature is between a and b degrees Centigrade.

As the reader has no doubt observed, we made many arbitrary choices in this CD. Among these are the very dissection of the world into situations and the ascription of temperature to each of these, the modeling of temperature by pairs of rationals (actually, we are getting ranges of temperature here), and the choice of scale in which to measure temperature. Our interest in this example lies not in the particular choices, but rather in the possibility that we might use the machinery of this chapter in giving an account of natural regularities and of how it is that agents are able to (sometimes) convey information to one another. One should see Barwise and Seligman [5] for a full exposition of the use of the machinery here in connection with the regularities of the natural world.

So now our motivations in this study are clearer. We would like to use CD's and whatever auxiliary apparatus is needed to study a range of phenomena involving classification and observation. In addition, there is the connection of this work to other studies about information. In connection with this last point, we might mention that several fields of mathematics are often thought of as *information driven*, and our study here can be thought of as an attempt to give conceptual foundations to these.

7.3   The Flow of Information

Classification-domains were introduced in Seligman [10] for the purpose of studying ideas about the flow of information. Information-flow can be regarded as a generalization of (Tarskian) logical consequence. If a model is classified by a sentence of a language, in the manner described in the previous section, then one can extract information about the model by making deductive inferences. For example, if the model is classified as being of type $A \vee (B \wedge C)$ then one can deduce that it is also of types $A \vee B$ and $A \vee C$, because both of these sentences are logical consequences of the first. This corresponds to the intuition that logical inference, a phenomenon which is internal to a CD and at the same time a universally applicable mode of information extraction, is more "fine-grained" than other types of information-flow.

In the general case, information sites are not analogous to classical models in that the information they contain may be partial and context-dependent. Information at one site may be used to extract information about a different site, either one which contains more extensive information than the first, or one which is contextually related to the first in some relevant way. In the terminology of classification domains, we want to characterize when one site s_1's being of a type t_1 "carries" the information that another site s_2 is of type t_2. When the information is carried, we say that the information *flows* from s_2 to s_1.

The terminology of "carrying information" and "information-flow" derive from Dretske's book [6], in which an information-based theory of knowledge is proposed. One of the main motivations for the study of classification domains was to provide a framework in which Dretske's ideas could be set aside more traditional ideas about information and logic. Whereas, in general, information flows between different sites, the deduction of logical consequences of a sentence can be regarded as information-flow within a single site. This is because a (Tarskian) consequence relation $\vdash$ is required to be *sound*: if $m \models \varphi$ and $\varphi \vdash \psi$ then $m \models \psi$.

The first attempt to generalize the soundness condition was given in by Barwise and Perry in [4] (and related papers) in connection with their development of Situation Semantics. They introduced *constraints* between pairs of (situation-)types. A constraint between types t_1 and t_2, written $t_1 \Rightarrow t_2$, was required to satisfy the condition that if any (actual) situation s_1 was of type t_1, there must exist a situation s_2 of type t_2.

For example, there is commonly thought to be a constraint between smoke and fire: if there is a lot of smoke around then it is likely that there is a fire nearby. Barwise and Perry account for this fact by postulating a constraint Sm$\Rightarrow$Fi between the type Sm of situations in which there is a lot of smoke around, and the type Fi of situations in which there is a fire burning. When one observes a lot of smoke, it is reasonable to conclude that the (observed) situation is of type Sm; then one can deduce that there is a situation of type Fi, a situation in which there is a fire burning. The example takes a few liberties with the notion of "situation," but it succeeds in making the point.

Barwise and Perry proposed to model many of the regularities of the world—including the "conventional" regularities that underlie linguistic communication—as constraints between types.[1] "Logical" laws can be modeled as constraints which obey the stronger condition that the situation of the antecedent type and the situation of the consequent type are the same.

Abstracting from the details of Barwise and Perry's Situation Semantics, Seligman (in [9]) introduced the notion of a *perspective*: a classification-domain together with a relation $\Rightarrow$ ("involves") between types obeying Barwise and Perry's condition.[2] The purpose of this abstraction was both to relativize constraints to a classification domain and to provide an analysis of the "structure" of types in terms of information flow.[3] Using perspectives, one can compare a variety of models of information flow—from the "logical" perspectives of classical, intuitionistic and modal logics, to the more "informational" perspectives based on probability theory, derived from Dretske's theory of knowledge.

However, there is something missing from Barwise and Perry's condition on $\Rightarrow$. Using existential quantification over sites is a very weak way of expressing information-flow. One would really like to know *which* sites are related in such a way that information at one can flow to information at the other. Even when considering natural regularities, like the relationship between smoke and fire, the quantifier is too weak. The presence of smoke indicates that there is a fire nearby because the smoke is usually caused by a particular fire; so there is a definite situation of type Fi, which is related to the observed one in the right way, although the observer may not be able to tell which one it is.

It is possible to define such a relation an obvious way: $s_1 \rightsquigarrow s_2$ iff for all types t_1 and t_2, if $s_1 \models t_1$ and $t_1 \Rightarrow t_2$ then $s_2 \models t_2$. This was done in [9]. But an alternative

---

[1] As well as their role in explaining the relationship between language and the world, constraints served the semantic purpose of being the referents of conditional sentences. This idea was developed by Barwise in [1] and by Barwise and Seligman in [5].

[2] In fact, perspectives have another relation $\perp$ ("precludes") between types obeying the condition that if s is of type t_1 and $t_1 \perp t_2$ then s is not of type t_2. This generalizes logical inconsistency in the same way that $\Rightarrow$ generalizes logical consequence. However, we ignore this aspect of perspectives for the sake of simplicity.

[3] For example, a logical type $A \wedge B$ has the structure of a conjunction in part because one can deduce from it both A and B.

approach, followed in [11], is to take relations between situations (called *channels* after Dretske's use of the term "communication channel") to be the primary object of study. According to this approach, what information flows from where to where depends on which channels are available. One can *define* involvement between types by considering which situations are related by the channel: $t_1 \Rightarrow t_2$ iff for all situations s_1 and s_2 related by the channel, if s_1 is of type t_1 then s_2 is of type t_2.

Finally, in [2], Barwise suggested that one needs *both* a relation between types and a relation between sites to give an adequate account of information flow. The notion of a perspective can then be generalized to that of an *information link* between (possibly different) classification domains.

7.4   Information Links between CD's

DEFINITION 7.1   Let $\langle S_1, \models_1, T_1 \rangle$ and $\langle S_2, \models_2, T_2 \rangle$ be CDs. A *pair of relations* between these two is a pair $\langle \leadsto, \Rightarrow \rangle$ consisting of an *indicating relation* between S_1 and S_2, together with a *signaling relation* between T_1 and T_2. A **(sound) link** between $\langle S_1, \models_1, T_1 \rangle$ and $\langle S_2, \models_2, T_2 \rangle$ is a pair of relations which satisfies the following condition:

If $t_1 \Rightarrow t_2$, then for all s_1 of type t_1, and all s_2 such that $s_1 \leadsto s_2$, s_2 must be of type t_2.

A **complete link** between $\langle S_1, \models_1, T_1 \rangle$ and $\langle S_2, \models_2, T_2 \rangle$ is a link which also satisfies the converse condition:

Suppose that whenever s_1 is of type t_1 and $s_1 \leadsto s_2$, then s_2 is of type t_2. Then $t_1 \Rightarrow t_2$.

The idea is that a sound link should allow us to learn properties of the target site (s_2) from any site which signals it. The exact information that we would learn depends is that s_2 is of type t_2, and we would only learn this if the source s_1 was of some type t_1 which indicated this.

A simple example comes from considering two logic CD's based on an inclusion of languages $\mathcal{L}_1 \subseteq \mathcal{L}_2$. This induces an indicating relation (in this case a function) $\Rightarrow$ in the obvious way, and we also have a natural signaling relation $\leadsto$ given by $A \leadsto B$ iff A is the $\mathcal{L}_1$-reduct of B. It is easy to check that this is a link. This squares with the intuitions above. This link is not complete, however, because distinct structures (even non-isomorphic structures) might satisfy the same formulas. This example also extends to the case of interpreting one language in another.

For another example, consider two topological spaces $\mathcal{T}_1$ and $\mathcal{T}_2$. Recall that a continuous function $f : \mathcal{T}_1 \to \mathcal{T}_2$ is a function from X_1 to X_2 so that for all $O \in \mathcal{O}_2$,

$f^{-1}(O) \in \mathcal{O}_1$. We construe f as the indicating part $\Rightarrow$ of a link, and we take for the signaling part the relation $\leadsto_f$ given by

$$X \leadsto_f Y \quad \text{iff} \quad f(X) \subseteq Y.$$

It is easy to check that $\langle f, \leadsto_f \rangle$ is a complete link: both soundness and completeness follow immediately from the definitions of the relations involved in the link. Moreover, the continuity of f was not used, so this example shows that a map between subset spaces (see Section 7.7 may be construed as a link.

The is a good place to mention that our definition of a sound link is not the only possibility. Indeed, our definition here differs from the one presented in Barwise [2]. Barwise considers a number of additional concepts; some of these have the flavor of the definition of constraint given in Section 7.3. One in particular is *co-absoluteness*: Whenever s_2 is in the range of the $\leadsto$ relation, if $s_2 \models t_2$, then there are s_1 and t_1 such that $s_1 \models t_1$ and $s_1 \leadsto s_2$. Returning to the discussion of continuous maps, we claim that f is continuous if and only if $\langle f, \leadsto_f \rangle$ is a complete and co-absolute link. The point is that f is continuous iff for each $x \in X_1$ and each $O_2 \in \mathcal{O}_2$ such that $f(x) \in O_2$, there is some O_1 containing x such that $f(O_1) \subseteq O_2$. This is easily seen to be equivalent to co-absoluteness.

One may regard the concept of information links as a formalization based on the metaphor of information exchange as a kind of liquid flow. After all, we speak of information *transfer, flow,* and *extraction.* So the definitions provide a mathematically precise way to think of the world as composed of a collection of CD's (or related structures), connected by links. On these links, information flows. (It should be mentioned that the natural composition operation on links is associative, so we can compose links.) This might be used, for example, in giving an account of how we come to know about the weather by reading thermometers. There is a link from the classification of parts of the world by the weather to classifications of situations involving thermometers by lenghts of mercury. (It is not so easy to spell out the exact details of this link, and we encourage the reader to try to do this.) There is also a link from this CD of thermometers to a CD of people and statments about the weather. The composition of these links would be a link from classifications of parts of the world by temperature to classifications of people by their knowledge or belief about the temperature. The goal of such an enterprise would be to give an account of information flow over "long distances" as a composition of "shorter" links.

A variation on this theme, one which we do not believe has been pursued, would be to propose *strengths of links* in order to give defaults for the breakdown of links. After all, if you walk outside today, read a thermometer and come to believe (temporarily) that it is $70°C$, you would probably suspect that the problem would be that the link between the weather and the thermometer was inoperative, due to a bad thermometer. You probably would not suspect your own powers of observation, that is, the link between you and the thermometer. What we mean by a breakdown of links here is that the links are given by conditional assertions, and the antecedents of those conditionals would simply not be true. For example, the first link would require that the thermometer actually works. In the case that it didn't, the link would not allow information flow. Then one might hope to use this factorization, for example in modeling some of our commonsense reasoning about the phenomenon.

7.5   Logic and Information Flow

Another approach to understanding information flow is more axiomatic in nature. One can propose plausible axioms and then look for interesting models.

In [3], Barwise suggests some postulates for the relation of "carrying" information when one is permitted to form the usual logical combinations of types. The relation has four arguments: the signal site, the target site, the indicating type, and the indicated type. We write s_1's being of type t_1 carries the information that s_2 is of type t_2 as

$$s_1 : t_1 \longrightarrow s_2 : t_2$$

Then Barwise's postulates can be stated as the following rules:

1. Xerox Principle (from [6]):

$$\frac{s_1 : t_1 \longrightarrow s_2 : t_2 \qquad s_2 : t_2 \longrightarrow s_3 : t_3}{s_1 : t_1 \longrightarrow s_3 : t_3}$$

2. Logic as Information Flow:

$$\frac{t_1 \vdash t_2}{s : t_1 \longrightarrow s : t_2}$$

3. Addition of Information:

$$s_1 : t_1 \longrightarrow s_2 : t_2 \quad s_1 : t_1' \longrightarrow s_2 : t_2'$$
$$s_1 : t_1 \wedge t_1' \longrightarrow s_2 : t_2 \wedge t_2'$$

4. Exhaustive Cases:

$$s_1 : t_1 \longrightarrow s_2 : t_2 \vee t_2' \quad s_2 : t_2 \longrightarrow s_3 : t_3 \quad s_2 : t_2' \longrightarrow s_3 : t_3$$
$$s_1 : t_1 \longrightarrow s_3 : t_3$$

5. Contraposition:

$$s_1 : t_1 \longrightarrow s_2 : t_2$$
$$s_2 : \neg t_2 \longrightarrow s_1 : \neg t_1$$

(In which, $\wedge$, $\vee$, $\neg$ and $\vdash$ are given a (classical) logical interpretation; e.g., $s \models t_1 \wedge t_2$ iff $s \models t_1$ and $s \models t_2$.) The reader should consult Barwise's paper for discussion of these principles.

One can define information carrying by an information link $\langle p, q \rangle$ by: $s_1 : t_1 \longrightarrow s_2 : t_2$ iff $s_1 \models t_1$, $p(s_1, s_2)$, $q(t_1, t_2)$ and $s_2 \models t_2$. Then, given a set of information links, one can say that information is carried just in case it is carried by one of the links in the set. Barwise's principles constitute closure conditions on the set of information links.

7.6 Channel Algebras

A complementary approach involves looking at the abstract properties of information links which are needed for a reasonable theory of information flow. Recent work by Barwise and Seligman has focussed on the following operations:

$c_1 \circ c_2$	sequential composition
$c_1 \wedge c_2$	parallel composition
c^{-1}	inverse
1	identity

To distinguish the elements of the abstract algebras from their concrete manifestations, the term "channel" is used instead of "information link". In the papers mentioned in this article the word "channel" has been used in various ways: in [11] channels relate situations; in [3] channels determine a signaling relation between sites, but are not identified with the relation they determine; and in [5] channels are classification domains of a special sort. One can think of each of these uses as concrete channel algebras:

channel algebra		
(i)	$c_1 \wedge c_2 = c_2 \wedge c_1$	$\wedge$ is commutative
(ii)	$c_1 \wedge (c_2 \wedge c_3) = (c_1 \wedge c_2) \wedge c_3$	$\wedge$ is associative
channel (iii)	$c \wedge c = c$	$\wedge$ is idempotent
algebra (iv)	$c_1 \circ (c_2 \circ c_3) = (c_1 \circ c_2) \circ c_3$	$\circ$ is associative
(v)	$1 \circ c = c \circ 1 = c$	1 is the unit of $\circ$
(vi)	$c_1 \circ (c_2 \wedge c_3) \leq (c_1 \circ c_2) \wedge (c_1 \circ c_3)$	left semi-distrib. of $\circ$ over $\wedge$
(vii)	$(c_2 \wedge c_3) \circ c_1 \leq (c_2 \circ c_1) \wedge (c_3 \circ c_1)$	right semi-distrib. of $\circ$ over $\wedge$

$$\vee$$

invertible channel algebra		
(viii)	$1^{-1} = 1$	1 is self-inverse
invertible (ix)	$c^{-1-1} = c$	$.^{-1}$ is an involution
channel (x)	$(c_1 \circ c_2)^{-1} = c_2^{-1} \circ c_1^{-1}$	inverse distrib. of $.^{-1}$ over $\circ$
algebra (xi)	$(c_1 \wedge c_2)^{-1} = c_1^{-1} \wedge c_2^{-1}$	distribution of $.^{-1}$ over $\wedge$
(xii)	$(c_1 \circ c_2) \wedge c_3 \leq (c_1 \wedge (c_3 \circ c_2^{-1})) \circ c_2$	modularity

Axioms (i) to (vii) define the class of *channel algebras*: structures with two binary operations, $\circ$ and $\wedge$, and a constant 1. Since $\wedge$ is commutative, associative and idempotent, it is the "meet" operation of the semi-lattice defined by: $c_1 \leq c_2$ iff $c_1 \wedge c_2 = c_1$. This ordering is used in stating axioms (vi), (vii) and (xii). Sequential composition, $\circ$, is a monoid with unit 1 which interacts with parallel composition by the semi-distributions (vi) and (vii).

Axioms (i) to (xii) define the class of *invertible channel algebras*: channel algebras with an inverse operation, $.^{-1}$, which in the relational case is just the operation of taking the converse.

These axioms are discussed in more detail in [3].

We will briefly consider what is involved in thinking of information-links as channel algebras. Given a classification domain C, let LINKC be the set of sound links betrween C and itself. How can we define operations on LINKC to make it into a channel algebra? The obvious choice is to use component-wise relational composition (;) for $\circ$, component-wise intersection for $\wedge$, and the identity link on C for 1.

These operations form a channel algebra on LINKC, but it has several undesirable properties. First, the operation of component-wise intersection is unnecessarily restrictive in the information it allows to flow. In particular, if some information is carried by $\langle p_1, q_1 \rangle$ it is not necessarily also carried by $\langle p_1 \cap p_2, q_1 \cap q_2 \rangle$. A better definition of parallel composition, which overcomes this problem, is given by: $\langle p_1, q_1 \rangle \circ \langle p_2, q_2 \rangle = \langle pp_1 \cap p_2, q_1 \cup q_2 \rangle$.

The second problem is more serious. There is no natural way of defining an inverse operation on LINKC. (The component-wise converse operation does not preserve soundness of links.) A solution to this problem is to define operations which simply maximize information flow. For each signaling relation p, define the indicating relation $\bar{p}$ (the *completion* of p) by: $\bar{p}(t_1, t_2)$ iff $\forall s_1, s_2 \in S$ if $s_1 \models t_1$ and $p(s_1, s_2)$ then $s_2 \models t_2$. Now note that, for any $p \subseteq S^2$, the pair $\langle p, \bar{p} \rangle$ is a sound and complete link on C. So the operations on links which maximize information flow are defined by:

$$\langle p_1, q_1 \rangle \circ \langle p_2, q_2 \rangle = \langle p_1; p_2, \overline{p_1; p_2} \rangle$$

$$\langle p_1, q_1 \rangle \wedge \langle p_2, q_2 \rangle = \langle p_1 \cap p_2, \overline{p_1 \cap p_2} \rangle$$

$$1 = \langle 1_S, \overline{1_S} \rangle$$

$$\langle p, q \rangle^{-1} = \langle p^{-1}, \overline{p^{-1}} \rangle$$

Now, these operations do not form a channel algebra on LINKC. In particular 1 is not a unit of $\circ$. This is because the operations "forget" all structure at the level of types. However, if we restrict ourselves to the set CLINK(C) of complete links between C and C, we get a channel algebra. (Every complete link is of the form $\langle p, \bar{p} \rangle$ for some p.)

Moreover, every subalgebra of CLINKC respects Barwise's postulates for information flow.[4]

7.7 Logic on Classification Domains

A somewhat different application of the basic idea of classification domains is to model the notions of knowledge and observation. The change is that now we think of the situations as being descriptions of the world, and we take the types to be observations that can be made. We do not know all of the properties of the actual situation, and we come to know these only through observations. Further, a stronger observation is one which cuts down the possible properties which a situation may have. So the inclusion order on types corresponds to strengthening of observations, or cutting down of possibilities. This leads to a logic for classification domains of Moss and Parikh [8], which we present in this section.

In addition to bringing to light a logic about CD's, a more fundamental purpose of this work is to attempt an analysis of different mathematical and computational fields as

---

[4]Results related to this are given in [3], but the presentation in terms of channel algebras and the relation to sound and complete links are new.

being *information driven* in some fashion. The point would be that fields such as general topology or recursive function theory can best be understood as studies based (implicitly) on something like an information-theoretic perspective. (Of course, it is not denied that, for example, topology is also about the intuitive ideas of smoothness and continuity. The point is that the axioms of point-set topology might correspond to something having to do with CD's, and that the elementary arguments of the subject might have more to do with that than with continuity.)

So far, work in this area has been mostly about general topology. Indeed, it is well appreciated that the elementary parts of this subject can be thought of as information-driven. (See, for example, Vickers [12] for an exposition of this view.) That is, the axioms concerning topologies can be motivated by thinking of the open sets as the records of possible observations.

Although we are primarily interested in the information-theoretic underpinnings of topology, we formulate our logical questions about a much larger class than the topological spaces.

DEFINITION 7.2   A **subset space** is a pair $\mathcal{X} = \langle X, \mathcal{O} \rangle$ where X is a set and $\mathcal{O}$ is a set of non-empty subsets of X. We assume that $X \in \mathcal{O}$, though this is really not necessary. $\mathcal{X}$ is **closed under intersection** if whenever $S, T \in \mathcal{O}$ and $S \cap T \neq \emptyset$, $S \cap T \in \mathcal{O}$. We can similarly define the notion of closure under *union*.

We now set up a formal language which is expressive enough for simple arguments concerning subset spaces.

DEFINITIONS   Let $\mathcal{A}$ be an arbitrary set of *atomic formulas*. $\mathcal{L}$ is the smallest set

containing each $A \in \mathcal{A}$, and closed under the following formation rules: if $\varphi, \psi \in \mathcal{L}$, then so are $\varphi \wedge \psi$ and $\neg\varphi$; if $\varphi \in \mathcal{L}$, then $K\varphi \in \mathcal{L}$ and $\Box\varphi \in \mathcal{L}$.

Let $\mathcal{X}$ be a subset space. Thinking of $\mathcal{X}$ as a Kripke *frame*, we give the semantics of $\mathcal{L}$ relative to an interpretation of the atomic formulas. Such an interpretation is a map $i : \mathcal{A} \to \mathcal{P}X$. The pair $(\mathcal{X}, i)$ will be called a *model*.

For $p \in X$ and $p \in u \in \mathcal{O}$, we define the **satisfaction relation** $\models_{\mathcal{M}}$ *on* $(X \times \mathcal{O}) \times \mathcal{L}$ by recursion on φ.

$p, u \models_{\mathcal{M}} A$ iff $p \in i(A)$
$p, u \models_{\mathcal{M}} \varphi \wedge \psi$ if $p, u \models \varphi$ and $p, u \models \psi$
$p, u \models_{\mathcal{M}} \neg\varphi$ if $p, u \not\models_{\mathcal{M}} \varphi$
$p, u \models_{\mathcal{M}} K\varphi$ if for all $q \in u$, $q, u \models_{\mathcal{M}} \varphi$
$p, u \models_{\mathcal{M}} \Box\varphi$ if for all $v \in \mathcal{O}$ such that $v \subseteq u$, $p, v \models_{\mathcal{M}} \varphi$

We adopt two abbreviations: $L\varphi$ means $\neg K \neg \varphi$, and $\diamond \varphi$ means $\neg \Box \neg \varphi$. So $p, u \models_{\mathcal{M}} L\varphi$ if there exists some $q \in u$ such that $q, u \models_{\mathcal{M}} \varphi$, and $p, u \models_{\mathcal{M}} \diamond \varphi$ if there exists $v \in \mathcal{O}$ such that $v \subseteq u$ and $p, v \models_{\mathcal{M}} \varphi$.

We write $p, u \models \varphi$ if $\mathcal{X}$ is clear from context. Finally, if $T \subseteq \mathcal{L}$, we write $T \models \varphi$ if for all models $\mathcal{M}$, all $p \in X$, and all $u \in \mathcal{O}$, if $p, u \models \psi$ for each $\psi \in T$, then also $p, u \models \varphi$.

If $\mathcal{X}$ is indeed a topology, then a set $i(A)$ will be open iff every point in $i(A)$ has an open neighborhood contained entirely in $i(A)$ iff for p in $i(A)$, the formula $\diamond K A$ holds. Thus $i(A)$ is *open* iff the formula $A \to \diamond K A$ is valid in the model. Dually, $i(A)$ is *closed* iff the formula $\Box L A \to A$ is valid in the model. We should mention that knowledge-theoretic intuitions are at work in other fields: with the obvious definitions, r.e. subsets of the natural numbers will satisfy the same knowledge theoretic formula that opens do in a topological setting, and this, we believe, is the source of the similarity. The set $i(A)$ is *dense* iff the formula LA is valid and it is *nowhere dense* if the formula $L\neg A$ is valid.

We axiomatize the $\models$ relation by taking as axioms all instances of the following formulas:

$$K\Box\varphi \to \Box K\varphi,$$

$$K\varphi \to (\varphi \wedge KK\varphi),$$

$$\varphi \to KL\varphi, \text{ and}$$

$$\Box\varphi \to (\varphi \wedge \Box\Box\varphi).$$

and we also take all instances of $A \to \Box A$ for A atomic. We use the following rules of inference: Modus Ponens, $\Box$-necessitation (from φ, deduce $\Box\varphi$), and K-necessitation (similarly).

THEOREM 7.1   Suppose $T \subseteq \mathcal{L}$. Then $T \vdash \varphi$ iff $T \models \varphi$.

The proof is too long to present here, mainly because there does not seem to be an easy presentation of a canonical model for this logic. That is, if we take the points of a model to be the maximal consistent theories in $\mathcal{L}$, there are no natural choices for the opens. This is also the case for the weaker completeness result for the same language with respect to the larger class of CD's: the only known proof involves a complicated construction. For a proof of this theorem, see [8].

More information about this logic can be found in [8] and the forthcoming Ph.D. thesis of Konstantinos Georgatos [7].

Once again, the point of studying this and other logics is to determine the extent to which elementary reasoning in a mathematical field like topology can be understood in terms of something simpler, such as a logic of this form. The claim is that some mathematical fields can be given an information-theoretic underpinning, and doing this would explain the naturalness of the axioms, and also the wide applicability of those fields.

References

[1]   Barwise, J. 1986. Conditionals and Conditional Information. Reprinted in Barwise, J. *The Situation in Logic*, CSLI Lecture Notes 17.

[2]   Barwise, J. 1991. Information Links in Domain Theory. To appear in the *Proceedings of the Conference on Mathematical Foundations of Program Semantics*, Springer LNCS.

[3]   Barwise, J. 1992. Constraints, Channels and Information Flow. To appear in *Situation Theory and its Applications*, vol III, CSLI Lecture Notes.

[4]   Barwise, J. and J. Perry. 1983. *Situations and Attitudes*, Bradford Books, MIT Press, Cambridge, Massachusetts.

[5]   Barwise, J. and J. Seligman. 1992. The Rights and Wrongs of Natural Regularity, to appear in J. Tomberlin (ed.) *Philosophical Perspectives*.

[6]   F. Dretske, F. 1981. *Knowledge and the Flow of Information*, Bradford Books, MIT Press, Cambridge, Massachusetts.

[7]   Georgatos, K. Forthcoming Ph.D. Thesis, City University of New York.

[8]   Moss, L., and Parikh, R. Topological Reasoning and the Logic of Knowledge, *Theoretical Aspects of Reasoning about Knowledge*, Y. Moses (ed.), 1992.

[9]   Seligman, J. 1990. Perspectives in Situation Theory. In R. Cooper, et al (eds), *Situation Theory and its Applications*, vol. I, CSLI Lecture Notes 22, 147–192.

[10] Seligman, J. 1990. *Perspectives: a relativistic approach to the theory of information*. Dissertation, Univ. of Edinburgh.

[11] Seligman, J. 1991. Physical Situations and Information Flow. In J. Barwise, et al. (eds), *Situation Theory and its Applications*, vol. II, CSLI Lecture Notes 26, 297–292.

[12] Vickers, S. 1989. *Topology via Logic*, Cambridge: Cambridge University Press.

8 Process Algebra and Dynamic Logic

Alban Ponse

8.1 Introduction

In this chapter, an extension of process algebra is introduced which can be compared to (propositional) dynamic logic. The additional feature is a 'guard' construct, related to the notion of a 'test' in dynamic logic. This extension of process algebra is semantically based on processes that transform data, and its operational semantics is defined relative to a structure describing these transformations via transitions between pairs of a process term and a data-state. The data-states are given by a structure that also defines in which data-states guards hold and how actions (non-deterministically) transform these states. The operational semantics is studied modulo strong bisimulation equivalence. For basic process algebra (without operators for parallelism) a small axiom system is presented which is complete with respect to a general class of data environments. In case a data environment satisfies some expressiveness constraints, (local) bisimilarity can be completely axiomatized by adding three axioms to this system.

Next, process algebra with parallelism and guards is introduced. A two-phase calculus is provided that makes it possible to prove identities between parallel processes. Also this calculus is complete. The use of the calculus is demonstrated by an extended example. The last section of the chapter consists of a short discussion on the operational meaning of the Kleene star operator.

The question how dynamic logic and process algebra can be integrated is intriguing. A well-known result relevant in this context stems from Hennessy and Milner, who defined a modal logic that characterizes observable equivalence between processes [7]. In process algebra with guards a different approach is followed, reminiscent of the semantical setting of Propositional Dynamic Logic (PDL). The present chapter outlines this approach.

In the generic sense 'process algebra' denotes an algebraic approach to the study of concurrent processes. Here a process is roughly "the behavior of a digital system", such as the execution of a computer program. The specific process theory introduced in this chapter is based on ACP (the Algebra of Communicating Processes), developed by Bergstra and Klop. For an overview of ACP see Baeten and Weijland in [2]. Other common algebraic concurrency theories are CCS (Calculus of Communicating Systems) developed by Milner [11] and CSP (Communicating Sequential Processes) overviewed by Hoare in [8].

Typical for the process algebra approach as followed in ACP is that one reasons with terms denoting processes, rather than with formulas expressing properties of a process (or a program) as is done in dynamic logic [6, 9]. This reasoning is equational, and the resulting identities refer to a *behavioral* equivalence: two processes are equal if they cannot be distinguished according to some notion of observability. A well-known behavioral equivalence is *bisimulation* [13], which is considered in this chapter. A small example: in ACP a standard axiom is

$$x + y = y + x$$

(the $+$ represents choice) which expresses that choice is independent of any ordering. A PDL interpretation of this axiom is

$$\langle x \cup y \rangle p \leftrightarrow \langle y \cup x \rangle p$$

the derivability of which depends on the PDL axiom $\langle x \cup y \rangle p \leftrightarrow \langle x \rangle p \vee \langle y \rangle p$ and propositional tautologies.

Another difference between dynamic logic and process algebra is that infinite behavior is a rather fundamental notion in the latter. A typical example of a (reactive) non terminating process is a communication protocol transmitting data through an unreliable channel, such that (despite the unreliability) no information will get lost. Correctness is then expressed recursively: its characteristic behavior is to transmit any received datum before a next datum can be received, i.e., the process behaves externally as a one-element buffer. Infinite processes can be defined as solutions of systems of recursive equations.

Process algebra with guards can be characterized by two starting points. The first typical ingredient is that processes are considered as having a separate *data-state*, contrary to the usual semantical setting in process algebra. The execution of a process is regarded in terms of atomic actions that transform data-states:

$$(a, s) \xrightarrow{a} (\epsilon, s')$$

where (a, s) represents the atomic action a in *initial* data-state s, the label a represents what single step can be observed, and $s' \in \mathit{effect}(a, s)$ is a data-state that can result from the execution of a (it is demanded that transformation sets of the form $\mathit{effect}(a, s)$ are not empty). The special constant ϵ ("empty process" or "**skip**") represents the possibility to terminate, and is associated with a termination transition characterized by the label $\sqrt{}$:

$$(\epsilon, s') \xrightarrow{\sqrt{}} (\delta, s')$$

where the special constant δ ("inaction" or "deadlock") indicates that no further activity can be performed. In the sequel a calculus is defined for deriving transitions from compound process terms, determining the operational semantics for process algebra with guards. As an example the process $a \cdot \delta$ (the $\cdot$ represents sequential composition) is able to perform an a-step to a configuration with process term $\epsilon \cdot \delta$ (which equals δ), that in turn allows no termination action. Bisimulation semantics can now be extended to a setting wherein data-states play an explicit role: e.g., the processes a and $a + a \cdot \epsilon$ are bisimilar if considered in *equal* initial date-states: each transition of the one process can be associated with (at least one) of the other process.

A second characteristic of process algebra with guards concerns the extension "with *guards*." Guards are comparable to tests in dynamic logic. Depending on the data-state, a guard can either be transparent such that it can be passed (so behaves like ϵ), or it can block and prevent subsequent processes from being executed (so behaves like δ). Typical for this extension is the one-sortedness: a guard *itself* represents a process. With this construct the guarded commands of DIJKSTRA [3] can be easily expressed, as well as a restricted notion of tests in PDL (in terms of [6] comparable with $PDL^{0.5}$, and so called *poor* tests in [9]). Guards have several nice properties, e.g., they constitute a Boolean algebra. Furthermore, a partial correctness formula

$$\{\alpha\}\, p\, \{\beta\}$$

can be expressed by the algebraic equation

$$\alpha\, p = \alpha\, p\, \beta$$

where α and β are guards (cf. [10]): this equation expresses that termination of the process p started in an initial data-state satisfying α, cannot be blocked by the postcondition β, i.e., β "holds" in this case. In the related paper [5] on process algebra with guards it is shown that Hoare logic for processes defined by linear recursion can be captured in a completely algebraic way (cf. [15]).

Parallel operators fit easily in the process algebra framework. In for instance Hoare logic, parallelism turns out to be rather intricate; proof rules for parallel operators are often substantial [12]. In the subsequent approach the difficulties caused by parallel operators in Hoare logic cannot be avoided, but can be dealt with in a simple algebraic way.

The chapter is organized as follows. Section 2 concerns a small fragment of process algebra with guards and introduces the fundamentals of the approach. A complete axiomatization of bisimilarity between finite processes with respect to a class of structures is presented, as well as an extended soundness result for processes defined by recursive equations. In Section 3 it is described that in case some expressiveness constraints are satisfied, bisimilarity relative to a single structure can also be completely characterized. Section 4 introduces the technical means to reason about parallel processes. These are illustrated in Section 5 by an extended example on the correctness of a parallel process. The chapter is concluded with a short discussion on the operational meaning of the Kleene star operator in Section 6. As to keep the chapter short, most proofs are omitted. However, all proofs are spelled out in [5].

8.2   Basic Process Algebra with Guards

Syntax and axioms.   Basic Process Algebra with guards, notation BPA_G, is parameterized by

1. A non-empty set A of *atomic actions*,

2. A non-empty set G_{at} of *atomic guards* disjunct from $A \cup \{\delta, \epsilon\}$.

Before defining the exact signature of BPA$_G$, the set of atomic guards is extended to the set G of *basic* guards in the following way. The two constants δ ('deadlock' or 'inaction') and ϵ ('empty process' or '**skip**') are added to G_{at}, and G is obtained by closure under negation, so contains elements φ satisfying the BNF clause

$$\varphi \quad ::= \quad \delta \mid \epsilon \mid \neg\varphi \mid \psi \in G_{at}.$$

The signature of BPA$_G$, notation $\Sigma(\text{BPA}_G)$, is defined by constants $a, b, c, \ldots$ representing the elements of A and constants $\varphi, \psi, \ldots$ representing the elements of G. Furthermore it contains the binary operators $+$ (choice) and $\cdot$ (sequential composition). In term formation brackets and variables of a set $V = \{x, y, z, \ldots\}$ are used. The function symbol $\cdot$ is often left out, and brackets are omitted according to the convention that $\cdot$ binds stronger than $+$. Finally, letters $t, t', \ldots$ are used to denote open terms, and letters $p, q, \ldots$ denote closed terms representing processes.

The axioms in Table 8.1 and those of equational logic express the basic identities between terms over $\Sigma(\text{BPA}_G)$. This axiom system is called BPA$_G^4$. The axioms A1 – A9 are well-known in process algebra (BPA$_{\delta\epsilon}$, [2]). Observe that there is no symmetric variant of the distributive axiom A4: an axiom $x(y + z) = xy + xz$ derives with atomic actions a, b for x, z and the constant δ for y the equation

$$ab = a\delta + ab$$

(use A6), identifying the process ab which is deadlock free with one that can behave as $a\delta$. The axioms G1 – G4 describe the fundamental identities between guards. G1 and G2 express that a basic guard always behaves dually to its negation: φ holds in a data-state s iff $\neg\varphi$ does not and vice versa. The axiom G3 states that $+$ does not change the interpretation of a basic guard φ. It does not matter whether the choice is exercised before or after the evaluation of φ. In the last new axiom G4 the following shorthand is used:

$$x \subseteq y \stackrel{def}{=} x + y = y \qquad (\text{and } x \supseteq y \stackrel{def}{=} y \subseteq x)$$

(this notation is called *summand inclusion*). This axiom can be motivated as follows: a process $a(\varphi p + \neg\varphi q)$ behaves either like ap or aq, depending on whether φ or $\neg\varphi$ can be passed in the data-state resulting from the execution of a. As a consequence the process $a(\varphi p + \neg\varphi q)$ should be a provable summand of $ap + aq$. The atomicity of a in this axiom is necessary. If a is for instance replaced by the term ab, then after a has happened it can be that execution of b yields a data-state where φ holds *and* a data-state where $\neg\varphi$ holds. Hence $ab(\varphi p + \neg\varphi q)$ need not be a summand of $abp + abq$. Note that the axiom G4 is not derivable from the first three 'guard'-axioms. The superscript 4 in BPA_G^4 indicates that there are four axioms referring to guards. Not all of these are always considered. In particular the system BPA_G^3, containing all BPA_G^4-axioms except G4 will play a role.

A1	$x + y = y + x$	G1 $\quad \varphi \cdot \neg\varphi = \delta$
A2	$x + (y + z) = (x + y) + z$	G2 $\quad \varphi + \neg\varphi = \epsilon$
A3	$x + x = x$	G3 $\quad \varphi(x + y) = \varphi x + \varphi y$
A4	$(x + y)z = xz + yz$	
A5	$(xy)z = x(yz)$	
A6	$x + \delta = x$	
A7	$\delta x = \delta$	G4 $\quad a(\varphi x + \neg\varphi y) \subseteq ax + ay$
A8	$\epsilon x = x$	
A9	$x\epsilon = x$	

Table 8.1
The axioms of BPA_G^4 where $\varphi \in G$ and $a \in A$

Up till now only 'basic' and 'atomic' guards were introduced. *Guards* as such, with typical elements $\alpha, \beta, \ldots$ are defined as terms over $\Sigma(\mathrm{BPA}_G)$ that contain only basic guards and the sequential and choice operators. The Boolean operator $\neg$ on guards can be defined by the *abbreviations*

$$\neg(\alpha\beta) \qquad \text{for} \qquad \neg\alpha + \neg\beta$$
$$\neg(\alpha + \beta) \qquad \text{for} \qquad \neg\alpha\neg\beta.$$

For guards there is the following theorem (cf. [16]):

THEOREM 8.1   Let G_{at} be a set of atomic guards. BPA_G^3 $(= \mathrm{BPA}_G^4 \setminus \mathrm{G4})$ is an equational basis for the Boolean algebra $(G_{at}, +, \cdot, \neg)$. ∎

Specifying processes by recursive equations.

DEFINITION 8.1   A *recursive specification* $E = \{x = t_x \mid x \in V_E\}$ over the signature $\Sigma(\mathrm{BPA}_G)$ is a set of equations where V_E is a (possibly infinitely) set of (indexed) variables and t_x a term over $\Sigma(\mathrm{BPA}_G)$ such that the variables in t_x are also in V_E.

A *solution* of a recursive specification $E = \{x = t_x \mid x \in V_E\}$ is an interpretation of the variables in V_E as processes, such that the equations of E are satisfied. For instance the recursive specification $\{x = x\}$ has any process as a solution for x and $\{x = ax\}$

has the infinite process "a^ω" as a solution for x. The following syntactical restriction on recursive specifications turns out to enforce unique solutions:

DEFINITION 8.2   Let t be a term over the signature $\Sigma(\text{BPA}_G)$. An occurrence of a variable x in t is *guarded* iff t has a subterm of the form $a \cdot M$ with $a \in A \cup \{\delta\}$, and this x occurs in M. Let $E = \{x = t_x \,|\, x \in V_E\}$ be a recursive specification over $\Sigma(\text{BPA}_G)$. The specification E is *guarded* iff all occurrences of variables in the terms t_x are guarded.

Note that the property "guarded" of a recursive specification has nothing to do with the "guards" that form the main subject of this chapter.

Now the signature $\Sigma(\text{BPA}_G)_{\text{REC}}$, containing representations of infinite processes, is defined as follows:

DEFINITION 8.3   The signature $\Sigma(\text{BPA}_G)_{\text{REC}}$ is obtained by extending $\Sigma(\text{BPA}_G)$ in the following way: for each guarded specification $E = \{x = t_x \,|\, x \in V_E\}$ over $\Sigma(\text{BPA}_G)$ a set of constants $\{<x\,|\,E> \,|\, x \in V_E\}$ is added, where the construct $<x\,|\,E>$ denotes the x-component of a solution of E.

Some more notations: let $E = \{x = t_x \,|\, x \in V_E\}$ be a guarded specification over $\Sigma(\text{BPA}_G)$, and t some term over $\Sigma(\text{BPA}_G)_{\text{REC}}$. Then $<t\,|\,E>$ denotes the term in which each occurrence of a variable $x \in V_E$ in t is replaced by $<x\,|\,E>$, e.g. the expression $<aax\,|\,\{x = ax\}>$ denotes the term $aa<x\,|\,\{x = ax\}>$.

For the constants of the form $<x\,|\,E>$ there are two axioms in Table 8.2. In these axioms the letter E ranges over guarded specifications. The axiom REC states that the constant $<x\,|\,E>$ ($x \in V_E$) is a solution for the x-component of E, so expresses that each guarded recursive system has *at least* one solution for each of its (bounded) variables. The conditional rule RSP (Recursive Specification Principle) expresses that E has *at most* one solution for each of its variables: whenever one can find processes p_x ($x \in V_E$) satisfying the equations of E, notation $E(\vec{p_x})$, then $p_x = <x\,|\,E>$.

REC $<x\,|\,E> = <t_x\,|\,E>$ if $x = t_x \in E$ and E guarded

RSP $\dfrac{E(\vec{p_x})}{p_x = <x\,|\,E>}$ if $x \in V_E$ and E guarded

---

Table 8.2
Axioms for guarded recursive specifications.

Finally, a convenient notation is to abbreviate $<x\,|\,E>$ for $x \in V_E$ by X once E is fixed, and to represent E only by its REC instances. The following example shows all notations concerning recursively specified processes, and illustrates the use of REC and RSP.

EXAMPLE 8.1   Consider the guarded recursive specifications $E = \{x = ax\}$ and $E' = \{y = ayb\}$ over $\Sigma(\text{BPA}_G)$. So by the convention just introduced, E can be represented by $X = aX$. With REC and RSP (and the congruence properties of $=$) one can prove

$$\text{BPA}_G + \text{REC} + \text{RSP} \vdash X = Y.$$

First note that $Xb = aXb$ by REC, so $E(Xb)$ is derivable. Application of RSP yields

$$Xb = X. \tag{8.1}$$

Moreover, $Xb \overset{\text{REC}}{=} aXb \overset{(8.1)}{=} aXbb$, and hence $E'(Xb)$ is derivable. A second application of RSP yields $Xb = Y$. Combining this with (8.1) gives the desired result.
End Example 8.1

Semantics.   In the set-up of process algebra with guards, a process is considered as having a *data-state*: an atomic action is a (non-deterministic) data-state *transformer* and a guard is a *test* on data-states. The operational semantics is defined relative to a structure over A, G_{at} that defines these three components:

DEFINITION 8.4   A *data environment* $\mathcal{S} = \langle S, \textit{effect}, \textit{test} \rangle$ over a set A of atomic actions and a set G_{at} of atomic guards is specified by

- A non-empty set S of data-states,
- A function $\textit{effect} : S \times A \to 2^S \setminus \{\emptyset\}$,
- A predicate $\textit{test} \subseteq S \times G_{at}$.

Observe that the function *effect* defining the state transformations possibly introduces non-determinism in state transformations. The predicate *test* determines whether an atomic guard holds in some data-state. Whenever $(s, \varphi) \in \textit{test}$, this means that in data-state s the atomic guard φ may be passed. In order to interpret basic guards, the predicate *test* is extended in the obvious way:

- for all $s \in S$ it holds that $(s, \epsilon) \in \textit{test}$ and that $(s, \delta) \notin \textit{test}$,
- for all $s \in S$ and $\varphi \in G$ it holds that $(s, \neg\varphi) \in \textit{test}$ iff $(s, \varphi) \notin \textit{test}$.

Processes are provided with an operational semantics in the style of PLOTKIN [14]. The behavior of a process p is defined by transitions between *configurations*.

DEFINITION 8.5   Let S be a set of data-states. A *configuration* (p, s) *over* $(\Sigma(\text{BPA}_G), S)$ is a pair containing a closed term p over $\Sigma(\text{BPA}_G)$ and a data-state $s \in S$. The set of all configurations over $(\Sigma(\text{BPA}_G), S)$ is denoted by $C(\Sigma(\text{BPA}_G), S)$.

Let $A_{\surd} \overset{def}{=} A \cup \{\surd\}$. The transition relation

$$\longrightarrow_{\Sigma(\mathrm{BPA}_G)_{\mathrm{REC}},\mathcal{S}} \subseteq C(\Sigma(\mathrm{BPA}_G),S) \times A_{\sqrt{}} \times C(\Sigma(\mathrm{BPA}_G),S)$$

contains all transitions between the configurations over $(\Sigma(\mathrm{BPA}_G)_{\mathrm{REC}},S)$ that are derivable with the rules for φ, a, $+$, $\cdot$ and recursion in Table 8.5 (see Section 8.4). The $\sqrt{}$-transitions signal termination of a process. The possible behavior associated to a term p in initial data-state s is captured by all transitions reachable from (p,s) in $\longrightarrow_{\Sigma(\mathrm{BPA}_G)_{\mathrm{REC}},\mathcal{S}}$.

The operational behavior embodied by such transitions can be characterized by *bisimulation equivalence* [13]. But following the traditional approach in semantics based on data-state transformations, processes with different data-states in their configurations are not compared with each other. To that end the standard notion of bisimilarity is adapted as follows:

DEFINITION 8.6   Let $\mathcal{S}$ be a data environment with data-state space S. A binary relation $R \subseteq C(\Sigma(\mathrm{BPA}_G)_{\mathrm{REC}},S) \times C(\Sigma(\mathrm{BPA}_G)_{\mathrm{REC}},S)$ is an *$\mathcal{S}$-bisimulation* iff R satisfies the transfer property, i.e. for all $(p,s),(q,s) \in C(\Sigma(\mathrm{BPA}_G)_{\mathrm{REC}},S)$ with $(p,s)R(q,s)$:

1. Whenever $(p,s) \xrightarrow{a}_{\Sigma(\mathrm{BPA}_G)_{\mathrm{REC}},\mathcal{S}} (p',s')$ for some a and (p',s'), then, for some q', also $(q,s) \xrightarrow{a}_{\Sigma(\mathrm{BPA}_G)_{\mathrm{REC}},\mathcal{S}} (q',s')$ and $(p',s')R(q',s')$,

2. Whenever $(q,s) \xrightarrow{a}_{\Sigma(\mathrm{BPA}_G)_{\mathrm{REC}},\mathcal{S}} (q',s')$ for some a and (q',s'), then, for some p', also $(p,s) \xrightarrow{a}_{\Sigma(\mathrm{BPA}_G)_{\mathrm{REC}},\mathcal{S}} (p',s')$ and $(p',s')R(q',s')$.

Two closed terms p,q over $\Sigma(\mathrm{BPA}_G)_{\mathrm{REC}}$ are *$\mathcal{S}$-bisimilar*, notation $p \underline{\leftrightarrow}_{\mathcal{S}} q$, iff for all $s \in S$ there is some $\mathcal{S}$-bisimulation R such that $(p,s)R(q,s)$.

The following lemma allows reasoning about bisimilarity in an algebraic way, and is crucial for the next two theorems.

LEMMA 8.1   For any data environment $\mathcal{S}$ the relation $\underline{\leftrightarrow}_{\mathcal{S}}$ between closed terms over $\Sigma(\mathrm{BPA}_G)_{\mathrm{REC}}$ is a congruence w.r.t. the operators of $\Sigma(\mathrm{BPA}_G)$.   ∎

THEOREM 8.2   *(Soundness)* Let p,q be closed terms over $\Sigma(\mathrm{BPA}_G)_{\mathrm{REC}}$. If $\mathrm{BPA}_G^4 + \mathrm{REC} + \mathrm{RSP} \vdash p = q$, then $p \underline{\leftrightarrow}_{\mathcal{S}} q$ for any data environment $\mathcal{S}$.   ∎

THEOREM 8.3   *(Completeness)* Let r_1, r_2 be closed terms over $\Sigma(\mathrm{BPA}_G)$. If $r_1 \underline{\leftrightarrow}_{\mathcal{S}} r_2$ for all data environments $\mathcal{S}$, then $\mathrm{BPA}_G^4 \vdash r_1 = r_2$.   ∎

8.3   BPA$_G$ in a Specific Data Environment

In this section bisimulation semantics for $\Sigma(\mathrm{BPA}_G)_{\mathrm{REC}}$ in a *specific* data environment is investigated.

A1	$x + y = y + x$	G1	$\varphi \cdot \neg\varphi = \delta$
A2	$x + (y + z) = (x + y) + z$	G2	$\varphi + \neg\varphi = \epsilon$
A3	$x + x = x$	G3	$\varphi(x + y) = \varphi x + \varphi y$
A4	$(x + y)z = xz + yz$	G4	$a(\varphi x + \neg\varphi y) \subseteq ax + ay$
A5	$(xy)z = x(yz)$		
A6	$x + \delta = x$	SI	$\varphi_0 \cdot \ldots \cdot \varphi_n = \delta$
A7	$\delta x = \delta$		if $\forall s \in S\ \exists i \leq n\ .\ (s, \varphi_i) \notin test$
A8	$\epsilon x = x$	WPC1	$wp(a, \varphi)a\varphi = wp(a, \varphi)a$
A9	$x\epsilon = x$	WPC2	$\neg wp(a, \varphi)a\neg\varphi = \neg wp(a, \varphi)a$

Table 8.3
The axioms of $\mathrm{BPA}_G(\mathcal{S})$ where $\varphi, \varphi_i \in G$ and $a \in A$.

In Table 8.3 the axiom system $\mathrm{BPA}_G(\mathcal{S})$ is presented. It contains the axioms of BPA_G^4 and three new axioms depending on $\mathcal{S}$ (this explains the $\mathcal{S}$ in $\mathrm{BPA}_G(\mathcal{S})$). The axiom SI (Sequence is Inaction) expresses that if a sequence of basic guards fails in each data-state, then it equals δ. Note that G1 follows from SI.

In the axioms WPC1 and WPC2 (Weakest Precondition under some Constraints) the expression $wp(a, \varphi)$ represents the basic guard that is the *weakest precondition* of an atomic action a and an atomic guard φ. Weakest preconditions are semantically defined as follows:

DEFINITION 8.7 Let A be a set of atomic actions, G_{at} a set of atomic guards and $\mathcal{S} = \langle S, \textit{effect}, \textit{test} \rangle$ be a data environment over A and G_{at}. A *weakest precondition* of an atomic action $a \in A$ and an atomic guard $\varphi \in G_{at}$ is a basic guard $\psi \in G$ satisfying for all $s \in S$:

$$test(\psi, s) \text{ iff } \forall s' \in S\ (s' \in effect(a, s) \Longrightarrow test(\varphi, s')).$$

If ψ is a weakest precondition of a and φ, it is denoted by $wp(a, \varphi)$. Weakest preconditions are *expressible* with respect to A, G_{at} and $\mathcal{S}$ iff there is a weakest precondition in G of any $a \in A$ and $\varphi \in G_{at}$.

DEFINITION 8.8 Let A be a set of atomic actions and G_{at} a set of atomic guards and let $\mathcal{S} = \langle S, \textit{effect}, \textit{test} \rangle$ be a data environment over A and G_{at}. The data environment $\mathcal{S}$ is *sufficiently deterministic* iff for all $a \in A$ and $\varphi \in G_{at}$:

$$\forall s, s', s'' \in S\ (s', s'' \in effect(a, s) \implies (test(\varphi, s') \Longleftrightarrow test(\varphi, s''))).$$

Remark that a data environment S with a deterministic function *effect* is sufficiently deterministic. If S is a data environment such that weakest preconditions are expressible and that is sufficiently deterministic then the axioms WPC1 and WPC2 exactly characterize the weakest preconditions in an algebraic way: WPC1 expresses that $wp(a, \varphi)$ is a precondition of a and φ and WPC2 states that $wp(a, \varphi)$ is the *weakest* precondition of a and φ. If on the other hand weakest preconditions are expressible in S, then the soundness of $\text{BPA}_G(S)$ implies that S is also sufficiently deterministic. With the axioms for weakest preconditions G4 becomes derivable, so both axioms G1 and G4 need not be considered in the following results characterizing $\underline{\leftrightarrow}_S$.

THEOREM 8.4   *(Soundness)* Let S be a data environment such that weakest preconditions are expressible and that is sufficiently deterministic. Let r_1, r_2 be closed terms over $\Sigma(\text{BPA}_G)_{\text{REC}}$. If $\text{BPA}_G(S) + \text{REC} + \text{RSP} \vdash r_1 = r_2$ then $r_1 \underline{\leftrightarrow}_S r_2$.   ∎

THEOREM 8.5   *(Completeness)* Let S be a data environment such that weakest preconditions are expressible and that is sufficiently deterministic. Let r_1, r_2 be closed terms over $\Sigma(\text{BPA}_G)$. If $r_1 \underline{\leftrightarrow}_S r_2$ then $\text{BPA}_G(S) \vdash r_1 = r_2$.   ∎

EXAMPLE 8.2   Process algebra with guards can be used to express and prove partial correctness formulas in Hoare logic. In [5] the soundness of a Hoare logic for process terms (see also [15]) is proved. Here a simple example that is often used as an illustration of Hoare logic is presented and its correctness is shown.

Let $\text{BPA}_G(S)$ represent a small programming language with Boolean guards and assignments. The language has the signature of $\Sigma(\text{BPA}_G)$ and further assume a set $V = \{x, y, ...\}$ of data variables. Actions have the form

$$[x := e]$$

with $x \in V$ a variable ranging over the integers $\mathbb{Z}$ and e an integer expression. Let some interpretation $[\![\cdot]\!]$ from closed integer expressions to integers be given. Atomic guards have the form

$$\langle e = f \rangle$$

where e and f are both integer expressions.

The components of the data environment $S = \langle S, \textit{effect}, \textit{test} \rangle$ are defined by:

1. $S = \mathbb{Z}^V$, i.e., the set of mappings from V to the integers, with typical element ρ;

2. $\textit{effect}([x := e], \rho) = \rho[[\![\rho(e)]\!]/x]$, assuming that the domain of ρ is extended to integer expressions in the standard way, and $\rho[n/x]$ is as the mapping ρ, except that x is mapped to n;

3. $\textit{test}(\langle e = f \rangle, \rho) \iff ([\![\rho(e)]\!] = [\![\rho(f)]\!])$.

Note that the effect function is deterministic, so S is certainly sufficiently deterministic. Weakest preconditions can easily be expressed:

$$wp([x := e], \langle e_1 = e_2 \rangle) = \langle e_1[e/x] = e_2[e/x] \rangle.$$

The axiom SI cannot be formulated so easily, partly because integer expressions are not yet defined very precisely. However, it can be characterized by the scheme:

$$\langle e_0 = f_0 \rangle \cdot ... \cdot \langle e_n = f_n \rangle = \delta \quad \text{iff} \quad \forall \rho \in S \; \exists i \leq n \; . \; [\![\rho(e_i)]\!] \neq [\![\rho(f_i)]\!].$$

Consider the following tiny program $SWAP$ that exchanges the initial values of x and y without using any other variables.

$$SWAP \quad \equiv \quad [x := x + y] \cdot [y := x - y] \cdot [x := x - y].$$

The correctness of this program can be expressed by the following equation:

$$\langle x = n \rangle \cdot \langle y = m \rangle \cdot SWAP$$
$$=$$
$$\langle x = n \rangle \cdot \langle y = m \rangle \cdot SWAP \cdot \langle x = m \rangle \cdot \langle y = n \rangle.$$

This equation says that if $SWAP$ is executed in an initial data-state where $x = n$ and $y = m$, then after termination of $SWAP$ it must hold, i.e. it can be derived, that $x = m$ and $y = n$. So $SWAP$ indeed exchanges the values of x and y.

The correctness of $SWAP$ can be proved as follows:

$$\langle x = n \rangle \cdot \langle y = m \rangle \cdot SWAP$$
$$\overset{\text{SI}}{=} \quad \langle (x + y) - y = n \rangle \cdot \langle (x + y) - ((x + y) - y) = m \rangle \cdot SWAP$$
$$\overset{\text{SI,WPC1}}{=} \quad \langle x = n \rangle \cdot \langle y = m \rangle \cdot [x := x + y] \cdot \langle x - y = n \rangle \cdot$$
$$\langle x - (x - y) = m \rangle \cdot [y := x - y] \cdot [x := x - y]$$
$$\overset{\text{WPC1}}{=} \quad \langle x = n \rangle \cdot \langle y = m \rangle \cdot [x := x + y] \cdot \langle x = n \rangle \cdot \langle y = m \rangle \cdot$$
$$[y := x - y] \cdot \langle y = n \rangle \cdot \langle x - y = m \rangle \cdot [x := x - y]$$
$$\overset{\text{WPC1}}{=} \quad \langle x = n \rangle \cdot \langle y = m \rangle \cdot SWAP \cdot \langle x = m \rangle \cdot \langle y = n \rangle.$$

End Example 8.2

8.4   Parallel Processes and Guards

The language of $\Sigma(BPA_G)$ is extended to $\Sigma(ACP_G)$ by adding the following four operators [1, 2]: the *encapsulation operator* ∂_H, the *merge* $\|$, the *left-merge* $\underline{\|}$ and the *communication-merge* |, suitable to describe the behavior of parallel, communicating processes. Encapsulation is used to enforce communication between processes. Communication is modeled by a communication function $\gamma : A \times A \longrightarrow A_\delta$ that is commutative and associative. If $\gamma(a, b)$ is δ, then a and b cannot communicate, and if $\gamma(a, b) = c$, then c is the action resulting from the communication between a and b. All general definitions for $\Sigma(\mathrm{BPA}_G)$ carry over to $\Sigma(\mathrm{ACP}_G)$, especially, $\Sigma(\mathrm{ACP}_G)_{\mathrm{REC}}$ denotes $\Sigma(\mathrm{ACP}_G)$ extended with all constants denoting solutions of guarded recursive specifications over $\Sigma(\mathrm{ACP}_G)$.

In Table 8.4 the axiom system ACP_G is presented (note that the axiom G4 is absent). Most of these axioms are standard for ACP and, apart from G1, G2 and G3, only the axioms EM10, EM11 and D0 are new. The axiom EM10 (EM11) expresses that a basic guard φ in $\varphi x \,\|\, y$ ($\varphi x \mid y$) may prevent both x and y from happening.

Using ACP_G any closed term over $\Sigma(\mathrm{ACP}_G)$ can be proved equal to one without merge operators, i.e., a closed term over $\Sigma(\mathrm{BPA}_G)$, by structural induction.

THEOREM 8.6   *(Elimination)* Let p be a closed term over $\Sigma(\mathrm{ACP}_G)$. There is a closed term q over $\Sigma(\mathrm{BPA}_G)$ such that $\mathrm{ACP}_G \vdash p = q$.   ∎

ACP_G and BPA_G^4 or $\mathrm{BPA}_G(\mathcal{S})$ cannot be combined in bisimulation semantics as $\underleftrightarrow{}_{\mathcal{S}}$ is not a congruence for the merge operators; if G4 is added to ACP_G one can derive

$$\mathrm{ACP}_G + \mathrm{G4} \quad \vdash \quad a(b \,\|\, d) + a(c \,\|\, d) + d(ab + ac) \tag{8.1}$$
$$= \quad (ab + ac) \,\|\, d$$
$$\overset{\mathrm{G4}}{=} \quad (ab + ac + a(\varphi b + \neg \varphi c)) \,\|\, d$$
$$\supseteq \quad a(\varphi b d + \neg \varphi c d + d(\varphi b + \neg \varphi c)). \tag{8.2}$$

So, in (2) it can be the case that after an a step φ holds, and a state is entered where a b or a d step can be performed. Performing the d step may yield a state were $\neg \varphi$ holds, so the only possible step left is a c step. This situation cannot be mimicked in (1): the only possible execution of adc in (1) has no b option after the a-step. Therefore, every term with (2) as a summand is not bisimilar to (1) for any reasonable form of bisimulation. So $\mathrm{ACP}_G + \mathrm{G4}$ is not sound in any bisimulation semantics.

As is it still the objective to prove $\mathcal{S}$-bisimilarity between closed terms containing merge operators, a *two-phase* calculus that does avoid these problems can be defined.

DEFINITION 8.9   *(A two-phase calculus $\vdash_2$)* Let p_1, p_2 be closed terms over $\Sigma(\mathrm{ACP}_G)_{\mathrm{REC}}$. Write

$$\mathrm{ACP}_G^4 \vdash_2 p_1 = p_2$$

iff there are closed terms q_1, q_2 over $\Sigma(\mathrm{BPA}_G)_{\mathrm{REC}}$ such that

$\text{ACP}_G \vdash p_i = q_i \quad (i = 1, 2),$
$\text{BPA}_G^4 \vdash q_1 = q_2.$

Furthermore, write

$\text{ACP}_G(\mathcal{S}) \vdash_2 p_1 = p_2$

iff there are closed terms q_1, q_2 over $\Sigma(\text{BPA}_G)_{\text{REC}}$ such that

$\text{ACP}_G \vdash p_i = q_i \quad (i = 1, 2),$
$\text{BPA}_G(\mathcal{S}) \vdash q_1 = q_2.$

Writing REC + RSP in front of $\vdash_2$ indicates that REC and RSP may be used in proving $p_i = q_i$ $(i = 1, 2)$ and $q_1 = q_2$.

Let $\mathcal{S} = \langle S, \text{effect, test} \rangle$ be some data environment over a set A of atomic actions and a set G_{at} of atomic guards. Table 8.5 contains the transition rules defining an operational semantics for $\Sigma(\text{ACP}_G)_{\text{REC}}$. Let

$$\longrightarrow_{\Sigma(\text{ACP}_G)_{\text{REC}}, \mathcal{S}} \subseteq C(\Sigma(\text{ACP}_G)_{\text{REC}}, S) \times A_\checkmark \times C(\Sigma(\text{ACP}_G)_{\text{REC}}, S)$$

be the transition relation containing all transitions that are derivable by these rules. The following definition introduces a different bisimulation equivalence, called *global $\mathcal{S}$-bisimilarity*, that is a congruence for the merge operators. The idea behind a global $\mathcal{S}$-bisimulation is that a context $p \parallel (.)$ around a process q can change the data-state of q at any time and global $\mathcal{S}$-bisimulation equivalence must be resistant against such changes. So, a configuration (p_1, s) is related to a configuration (p_2, s) if $(p_1, s) \xrightarrow{a} (q_1, s')$ implies $(p_2, s) \xrightarrow{a} (q_2, s')$ and, as the environment may change s', q_1 is related to q_2 in *any* data-state:

DEFINITION 8.10   Let $\mathcal{S}$ be a data environment with data-state space S. A binary relation

$$R \subseteq C(\Sigma(\text{ACP}_G)_{\text{REC}}, S) \times C(\Sigma(\text{ACP}_G)_{\text{REC}}, S)$$

is a *global $\mathcal{S}$-bisimulation* iff R satisfies the following (global) version of the transfer property: for all $(p, s), (q, s) \in C(\Sigma(\text{ACP}_G)_{\text{REC}}, S)$ with $(p, s)R(q, s)$:

1.   Whenever $(p, s) \xrightarrow{a}_{\Sigma(\text{ACP}_G)_{\text{REC}}, \mathcal{S}} (p', s')$ for some a and (p', s'), then, for some q', also $(q, s) \xrightarrow{a}_{\Sigma(\text{ACP}_G)_{\text{REC}}, \mathcal{S}} (q', s')$ and $\forall s \in S \, ((p', s)R(q', s))$,

2.   Whenever $(q, s) \xrightarrow{a}_{\Sigma(\text{ACP}_G)_{\text{REC}}, \mathcal{S}} (q', s')$ for some a and (q', s'), then, for some p', also $(p, s) \xrightarrow{a}_{\Sigma(\text{ACP}_G)_{\text{REC}}, \mathcal{S}} (p', s')$ and $\forall s \in S \, ((p', s)R(q', s))$.

Two closed terms p, q over $\Sigma(\text{ACP}_G)_{\text{REC}}$ are *globally S-bisimilar*, notation

$$p \leftrightarrow_S q$$

iff for each $s \in S$ there is a global S-bisimulation relation R with $(p, s)R(q, s)$.

By definition of global S-bisimilarity it follows that

$$p \leftrightarrow_S q \implies p \leftrightarrow_S q$$

for closed terms p, q over $\Sigma(\text{ACP}_G)_{\text{REC}}$. Moreover, global S-bisimilarity *is* a congruence relation:

> LEMMA 8.2   For any data environment S the relation $\leftrightarrow_S$ is a congruence with respect to the operators of $\Sigma(\text{ACP}_G)$. ∎

> THEOREM 8.7   *(Soundness)* Let p, q be closed terms over $\Sigma(\text{ACP}_G)_{\text{REC}}$.
>
> 1. If $\text{ACP}_G + \text{REC} + \text{RSP} \vdash p = q$, then $p \leftrightarrow_S q$ for any data environment S.
> 2. If $\text{ACP}_G^4 + \text{REC} + \text{RSP} \vdash_2 p = q$, then $p \leftrightarrow_S q$ for any data environment S.
> 3. Let S be a data environment such that weakest preconditions are expressible and that is sufficiently deterministic.
> If $\text{ACP}_G(S) + \text{REC} + \text{RSP} \vdash_2 p = q$, then $p \leftrightarrow_S q$. ∎

> THEOREM 8.8   *(Completeness)* Let r_1, r_2 be closed terms over $\Sigma(\text{ACP}_G)$.
>
> 1. If $r_1 \leftrightarrow_S r_2$ for all data environments S, then $\text{ACP}_G \vdash r_1 = r_2$.
> 2. If $r_1 \leftrightarrow_S r_2$ for all data environments S, then $\text{ACP}_G^4 \vdash_2 r_1 = r_2$.
> 3. Let S be a data environment such that weakest preconditions are expressible and that is sufficiently deterministic.
> If $r_1 \leftrightarrow_S r_2$, then $\text{ACP}_G(S) \vdash_2 r_1 = r_2$. ∎

8.5   An Example: A Parallel Predicate Checker

In this section the techniques introduced up till now are illustrated by an example. Let $f \subseteq \mathbb{Z}$ be some predicate, e.g., the set of all primes. Now, given some number n, the objective is to calculate the smallest $m \geq n$ such that $f(m)$. Assume two devices P_1 and P_2 that can calculate for some given number k whether $f(k)$ holds. In Figure 8.1 a system is depicted that enables a calculation of m using both P_1 and P_2. A Generator/Collector G generates numbers $n, n+1, n+2, ...$, sends them to P_1 or P_2, and collects their answers. Furthermore G selects the smallest number satisfying f from the answers and presents it to the environment.

To describe this situation, Example 8.2 is extended with the atomic actions ($i = 1, 2$):

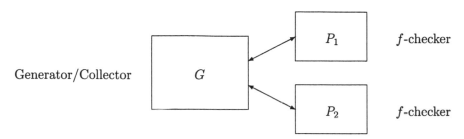

Generator/Collector

G

P_1 f-checker

P_2 f-checker

Figure 8.1
The parallel predicate checker Q.

$s(!x)$	send value of x,
$s_{ok}(!x_i)$	send the value x_i for which the evaluation of $f(x_i)$ was a success,
s_{notok}	indicate that an evaluation of f was not successful,
$r(?x_i)$	read a value for x_i,
$r_{ok}(?y)$	read a value for y for which $f(y)$ succeeded,
r_{notok}	read that an evaluation of f has failed,
c_{notok}	a communication between r_{notok} and s_{notok},
$w(!x)$, $w(!y)$	write value of x, y to environment.

These atomic actions communicate according to the following scheme:

$$\gamma(s(!x), r(?x_i)) = \gamma(r(?x_i), s(!x)) = [x_i := x],$$
$$\gamma(s_{ok}(!x_i), r_{ok}(?y)) = \gamma(r_{ok}(?y), s_{ok}(!x_i)) = [y := x_i],$$
$$\gamma(s_{notok}, r_{notok}) = \gamma(r_{notok}, s_{notok}) = c_{notok}.$$

All new atomic actions do not change the data-state, i.e. for each new atomic action a:

$$effect(a, \rho) = \{\rho\}.$$

Probably, one would expect that for instance $effect(r(?y), \rho) = \{\rho[new\ value/y]\}$ as $r(?y)$ reads a new value for y. But this need not be so: the value of y is only changed if a communication takes place.

Let new atomic guards $\langle f(t) \rangle$ for any integer expression t be added to the setting of Example 8.2. These guards have their obvious interpretation: $test(\langle f(t) \rangle, \rho)$ holds iff $f([\![\rho(t)]\!])$ holds.

The parallel predicate checker Q can now be specified by:

$$G = [x := n]\, s(!x)\, [x := x + 1]\, s(!x)\, G_1$$
$$G_1 = r_{notok}\, [x := x + 1]\, s(!x)\, G_1 + r_{ok}(?y)\, G_2$$
$$G_2 = \neg\langle x = y \rangle\, w(y) + \langle x = y \rangle (r_{ok}(?y)\, w(y) + r_{notok}\, w(x))$$

$$P_i = r(?x_i)\, P_i' + \epsilon$$
$$P_i' = \langle f(x_i) \rangle s_{ok}(!x_i) + \neg\langle f(x_i) \rangle s_{notok}\, P_i + \epsilon$$

$$Q = \partial_H(G \parallel (P_1 \parallel P_2))$$

with $H = \{r(?x_i), r_{ok}(?y), r_{notok}, s(!x), s_{ok}(!x_i), s_{notok} \mid i = 1, 2\}$.

The parallel predicate checker Q is correct if directly before the execution of an atomic action $w(x)$ or $w(y)$, x respectively y represents the smallest number $m \geq n$ such that $f(m)$. Let new atomic guards $\langle \alpha(t, u) \rangle$ for integer expressions t, u be of help to express this formally:

$$test(\langle \alpha(t, u) \rangle, \rho) \iff [\![\rho(t)]\!] \leq [\![\rho(u)]\!] \wedge \left(\bigwedge_{\substack{n \leq j < [\![\rho(u)]\!] \\ j \neq [\![\rho(t)]\!]}} \neg f(j) \right).$$

Now Q is correct if $\mathrm{ACP}_G(\mathcal{S}) + \mathrm{REC} + \mathrm{RSP} \vdash_2 Q = Q'$, where Q' is defined by:

$$Q' = \partial_H(G' \parallel (P_1 \parallel P_2))$$

with H, P_1 and P_2 as above, and G' is defined by (the difference between G and G' is underlined):

$$G' = [x := n]\, s(!x)\, [x := x + 1]\, s(!x)\, G_1'$$
$$G_1' = r_{notok}\, [x := x + 1]\, s(!x)\, G_1' + r_{ok}(?y)\, G_2'$$
$$\begin{aligned} G_2' = {} & \neg\langle x = y \rangle \cdot \underline{\langle \alpha(y, y) \rangle \langle f(y) \rangle} \cdot w(y) + \\ & \langle x = y \rangle (r_{ok}(?y) \cdot \underline{\langle \alpha(y, y) \rangle \langle f(y) \rangle} \cdot w(y) + \\ & \qquad r_{notok} \cdot \underline{\langle \alpha(x, x) \rangle \langle f(x) \rangle} \cdot w(x)). \end{aligned}$$

Note that α is unnecessarily complex to state the correctness of Q. But this formulation is useful in the second phase of the proof of $\mathrm{ACP}_G(\mathcal{S}) + \mathrm{REC} + \mathrm{RSP} \vdash_2 Q = Q'$.

This proof is given by first expanding Q and Q' to the *merge*-free forms R and R':

$$R \quad = \quad [x := n]([x_1 := x] [x := x + 1] [x_2 := x] \cdot R_1 +$$
$$[x_2 := x] [x := x + 1] [x_1 := x] \cdot R_1 \quad)$$

$$R_1 \quad = \quad \neg \langle f(x_1) \rangle c_{notok} [x := x + 1] [x_1 := x] \cdot R_1 +$$
$$\neg \langle f(x_2) \rangle c_{notok} [x := x + 1] [x_2 := x] \cdot R_1 +$$
$$\langle f(x_1) \rangle [y := x_1] R_2 +$$
$$\langle f(x_2) \rangle [y := x_2] R_3$$

$$R_2 \quad = \quad \neg \langle x = y \rangle w(y) +$$
$$\langle x = y \rangle (\langle f(x_2) \rangle [y := x_2] w(y) + \neg \langle f(x_2) \rangle c_{notok} w(x))$$

$$R_3 \quad = \quad \neg \langle x = y \rangle w(y) +$$
$$\langle x = y \rangle (\langle f(x_1) \rangle [y := x_1] w(y) + \neg \langle f(x_1) \rangle c_{notok} w(x)).$$

The process R' is defined likewise, except that

- $w(x)$ is replaced by $\langle \alpha(x, x) \rangle \langle f(x) \rangle w(x)$, and
- $w(y)$ is replaced by $\langle \alpha(y, y) \rangle \langle f(y) \rangle w(y)$.

It can be proved that

$$\text{ACP}_G + \text{REC} + \text{RSP} \vdash Q = R \qquad \text{and}$$
$$\text{ACP}_G + \text{REC} + \text{RSP} \vdash Q' = R'.$$

In order to show that $\text{BPA}_G(\mathcal{S}) + \text{REC} + \text{RSP} \vdash R = R'$ the following instances of SI, WPC1 and WPC2 are needed in addition to those given in Example 8.2. Let F be some function on integer expressions.

$$\varphi \, c_{notok} \, \varphi = \varphi \, c_{notok} \text{ for all } \varphi \in G,$$
$$\neg \langle t = t \rangle = \delta,$$
$$\langle t = u \rangle \neg \langle u = t \rangle = \delta,$$
$$\langle t = u \rangle \langle u = v \rangle \neg \langle t = v \rangle = \delta,$$
$$\langle t_1 = u_1 \rangle \cdot \ldots \cdot \langle t_k = u_k \rangle \neg \langle F(t_1, ..., t_k) = F(u_1, ..., u_k) \rangle = \delta,$$
$$\langle t + 1 = u \rangle \langle t = u \rangle = \delta,$$
$$\neg \langle f(t) \rangle \langle \alpha(t, u) \rangle \neg \langle \alpha(u, u + 1) \rangle = \delta,$$
$$\neg \langle f(t) \rangle \langle \alpha(u, t) \rangle \neg \langle \alpha(u, t + 1) \rangle = \delta,$$
$$\langle \alpha(t, u - 1) \rangle \langle t = u \rangle = \delta.$$

Note that these identities are valid. Let

$$\beta \stackrel{def}{=} \neg \langle x_1 = x_2 \rangle (\langle \alpha(x_1, x_2) \rangle \langle x = x_2 \rangle + \langle \alpha(x_2, x_1) \rangle \langle x = x_1 \rangle).$$

It is easy to show that

$$R \, , \quad \beta \, R_1 \, , \quad \langle y = x_1 \rangle \langle f(x_1) \rangle \beta \, R_2 \, , \quad \langle y = x_2 \rangle \langle f(x_2) \rangle \beta \, R_3$$

and

$$R' \, , \quad \beta \, R_1' \, , \quad \langle y = x_1 \rangle \langle f(x_1) \rangle \beta \, R_2' \, , \quad \langle y = x_2 \rangle \langle f(x_2) \rangle \beta \, R_3'$$

are solutions for T, T_1, T_2 and T_3, respectively, in the following specification:

$$
\begin{aligned}
T \;=\; & [x := n]([x_1 := x]\,[x := x+1]\,[x_2 := x] \cdot T_1 \;+ \\
& \quad [x_2 := x]\,[x := x+1]\,[x_1 := x] \cdot T_1 \quad)
\end{aligned}
$$

$$
\begin{aligned}
T_1 \;=\; & \beta\,(\neg\langle f(x_1)\rangle c_{notok}\,[x := x+1]\,[x_1 := x] \cdot T_1 \;+ \\
& \quad \neg\langle f(x_2)\rangle c_{notok}\,[x := x+1]\,[x_2 := x] \cdot T_1 \;+ \\
& \quad \langle f(x_1)\rangle\,[y := x_1]\,T_2 \;+ \\
& \quad \langle f(x_2)\rangle\,[y := x_2]\,T_3
\end{aligned}
$$

$$
\begin{aligned}
T_2 \;=\; & \langle y = x_1 \rangle \langle f(x_1) \rangle \beta(\neg\langle x = y \rangle\,w(y) \;+ \\
& \quad \langle x = y \rangle(\;\; \langle f(x_2)\rangle\,[y := x_2]\,w(y) \;+ \\
& \quad\quad\quad\quad \neg\langle f(x_2)\rangle\,c_{notok}\,w(x)))
\end{aligned}
$$

$$
\begin{aligned}
T_3 \;=\; & \langle y = x_2 \rangle \langle f(x_2) \rangle \beta(\neg\langle x = y \rangle\,w(y) \;+ \\
& \quad \langle x = y \rangle(\;\; \langle f(x_1)\rangle\,[y := x_1]\,w(y) \;+ \\
& \quad\quad\quad\quad \neg\langle f(x_1)\rangle\,c_{notok}\,w(x)))
\end{aligned}
$$

and thus $\mathrm{BPA}_G(\mathcal{S}) + \mathrm{REC} + \mathrm{RSP} \vdash R = R'$. Using (8.5) above it follows that

$$\mathrm{ACP}_G(\mathcal{S}) + \mathrm{REC} + \mathrm{RSP} \vdash_2 Q = Q'$$

as was to be proved.

8.6   Discussion on the Kleene Star

It is for the author still an open question whether the Kleene star operator $*$ can be incorporated in process algebra such that its operational meaning is captured, and such that bisimulation semantics can be characterized by (conditional) equational laws.

Following [4], a first alternative is to consider p^* as the process X defined by

$$X = pX + \epsilon$$

and therefore satisfying the axiom

$$x^* = x \cdot x^* + \epsilon$$

(which also implies what transition rules are appropriate). The obvious mismatch is that for e.g. an atomic action a the process a^* may give rise to an infinite execution, contrary to how it is often characterized. On the other hand, the conventional program **while** φ **do** a **od** would then translate in

$$(\varphi \cdot a)^* \neg \varphi \quad \text{(cf. [6, 9]) i.e.,} \quad Y \neg \varphi \quad \text{where} \quad Y = \varphi a Y + \epsilon$$

which is a deterministic process: $Y \neg \varphi$ performs from a data-state s an a-transition to $(Y \neg \varphi, \textit{effect}(s, a))$ iff $(s, \varphi) \in \textit{test}$, and a termination transition otherwise. Indeed, taking ϵ (i.e., **true**) for φ, this program represents an infinite a-loop, according to what should be expected.

A second alternative is to interpret for instance a^* as the process defined by

$$\epsilon + \sum_{0 < i < \omega} a^i.$$

One of the problems in this case is that it would require rather strong proof principles to distinguish this process from

$$\epsilon + \sum_{0 < i \leq \omega} a^i.$$

Adopting this third alternative again introduces an infinite a-branch. It should be remarked that the second alternative above matches most closely with the description of the * operator as given in [9]:

$p^* :=$ "Execute p repeatedly a nondeterministically chosen finite
       number of times."

Assume $\Sigma(\text{BPA}_G)$ is extended with the Kleene star * with its operational meaning according to the first alternative, i.e.,

$$p^* = p \cdot p^* + \epsilon.$$

Consider the following extension of BPA_G^4 with the five axioms K11 – K15 on the Kleene star (call the resulting system BPA_G^*):
The question whether these particular axioms completely characterize bisimilarity over $\Sigma(\text{BPA}_G^*)$ is a topic of further research. However, the following typical identities are derivable in BPA_G^*:

1. $\delta^* = \epsilon$ [Apply Kl1, A7 and A6],
2. $(x^*)^* = x^*$ [Apply Kl3 on $(\delta + x^*)^*$],
3. $x^* \cdot x^* = x^*$ [Apply Kl1 on $(x^*)^*$],
4. $\epsilon^* = \epsilon$ [Apply 1 and 2],
5. $(x + \epsilon)^* = x^*$ [Apply 1 and Kl3],
6. $(x + \epsilon)(x + y)^* = (x + y)^*$ [Apply Kl1 on $(x + y)^*$, A3, Kl1],
7. $x^*(x + y)^* = (x + y)^*$ [Apply Kl3, Kl1 on $(x + y)^*$],
8. $(x^*y^*)^* = (x + y)^*$ [Apply Kl4, Kl3, 7, omit 'ϵ+'],
9. $(x + y^*z^*)^* = (x + y + z)^*$ [Apply Kl3, 8, Kl3],
10. $(x^* \cdot y^* \cdot z^*)^* = (x + y + z)^*$ [Apply Kl4, 9, 7, omit 'ϵ+'],
11. $((x + \epsilon)(x + \epsilon))^* = x^*$ [Apply Kl1, 6 and omit '$+\epsilon$'].

As for the expressivity of $\Sigma(\mathrm{BPA}^*_G)$, it does not seem to be the case that all finitely branching *finite state* processes are specifiable (modulo bisimulation semantics). Consider for example the following transition system over a trivial data environment with singleton state space (its root is marked with a little arrow):

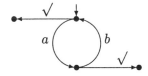

which seems not to be specifiable (the right-hand side termination step causes the problem: termination *within* an iteration can only occur (?) if at that state a new iteration exists). Observe that replacing b by a makes the transition system above specifiable, e.g., by a^*.

References

[1]   J.A. Bergstra and J.W. Klop. The algebra of recursively defined processes and the algebra of regular processes. In J. Paredaens, editor, *Proceedings* 11th *ICALP,* Antwerp, volume 172 of *Lecture Notes in Computer Science*, pages 82–95. Springer-Verlag, 1984.

[2]   J.C.M. Baeten and W.P. Weijland. *Process Algebra*. Cambridge Tracts in Theoretical Computer Science 18. Cambridge University Press, 1990.

[3]   E.W. Dijkstra. *A Discipline of Programming*. Prentice-Hall International, Englewood Cliffs, 1976.

[4]   R.T.P. Fernando. Parallelism, partial evaluation and programs as relations on states. In preparation.

[5]   J.F. Groote and A. Ponse. Process algebra with guards. Report CS-R9069, CWI, Amsterdam, 1990. To appear in *Formal Aspects of Computing*.

[6]   D. Harel. Dynamic logic. In D. Gabbay and F. Guenthner, editors, *Handbook of Philosophical Logic*, volume II, pages 497–604. Reidel, 1984.

[7]   M. Hennessy and R. Milner. Algebraic laws for nondeterminism and concurrency. *Journal of the ACM*, 32(1), 137–161, 1985.

[8]   C.A.R. Hoare. *Communicating Sequential Processes*. Prentice-Hall International, Englewood Cliffs, 1985.

[9]   D. Kozen and J. Tiuryn. Logics of programs. *Handbook of Theoretical Computer Science*, pages 789–840. Elsevier Science Publishers, 1990.

[10]   E.G. Manes and M.A. Arbib. *Algebraic Approaches to Program Semantics*. Texts and Monographs in Computer Science. Springer-Verlag, 1986.

[11]   R. Milner. *A Calculus of Communicating Systems*, volume 92 of *Lecture Notes in Computer Science*. Springer-Verlag, 1980.

[12]   S. Owicki and D. Gries. An axiomatic proof technique for parallel programs. *Acta Informatica*, pages 319–340, 1976.

[13]   D.M.R. Park. Concurrency and automata on infinite sequences. In P. Deussen, editor, 5th *GI Conference*, volume 104 of *Lecture Notes in Computer Science*, pages 167–183. Springer-Verlag, 1981.

[14]   G.D. Plotkin. A structural approach to operational semantics. Report DAIMI FN-19, Computer Science Department, Aarhus University, 1981.

[15]   A. Ponse. Process expressions and Hoare's logic. *Information and Computation*, 95(2):192–217, 1991.

[16]   F.M. Sioson. Equational bases of Boolean algebras. *Journal of Symbolic Logic*, 29(3):115–124, September 1964.

A1	$x + y = y + x$	G1	$\varphi \cdot \neg\varphi = \delta$
A2	$x + (y + z) = (x + y) + z$	G2	$\varphi + \neg\varphi = \epsilon$
A3	$x + x = x$	G3	$\varphi(x + y) = \varphi x + \varphi y$
A4	$(x + y)z = xz + yz$		
A5	$(xy)z = x(yz)$		
A6	$x + \delta = x$		
A7	$\delta x = \delta$		
A8	$\epsilon x = x$		
A9	$x\epsilon = x$		

CF	$a \mid b = \gamma(a, b)$

EM1	$x \parallel y = x \mathbin{\lfloor\!\lfloor} y + y \mathbin{\lfloor\!\lfloor} x + x \mid y$	EM10	$\varphi x \mathbin{\lfloor\!\lfloor} y = \varphi(x \mathbin{\lfloor\!\lfloor} y)$
EM2	$\epsilon \mathbin{\lfloor\!\lfloor} x = \delta$	EM11	$\varphi x \mid y = \varphi(x \mid y)$
EM3	$ax \mathbin{\lfloor\!\lfloor} y = a(x \parallel y)$		
EM4	$(x + y) \mathbin{\lfloor\!\lfloor} z = x \mathbin{\lfloor\!\lfloor} z + y \mathbin{\lfloor\!\lfloor} z$	D0	$\partial_H(\varphi) = \varphi$
EM5	$x \mid y = y \mid x$	D1	$\partial_H(a) = a$ if $a \notin H$
EM6	$\epsilon \mid \epsilon = \epsilon$	D2	$\partial_H(a) = \delta$ if $a \in H$
EM7	$\epsilon \mid ax = \delta$	D3	$\partial_H(x + y)$
EM8	$ax \mid by = (a \mid b)(x \parallel y)$		$= \partial_H(x) + \partial_H(y)$
EM9	$(x + y) \mid z = x \mid z + y \mid z$	D4	$\partial_H(xy) = \partial_H(x)\partial_H(y)$

Table 8.4
The axioms of ACP_G, $a, b \in A$, $H \subseteq A$ and $\varphi \in G$.

$\varphi \in G \quad (\varphi, s) \xrightarrow{\sqrt{}} (\delta, s) \quad$ if $test(\varphi, s)$

$a \in A \quad (a, s) \xrightarrow{a} (\epsilon, s') \quad$ if $s' \in effect(a, s)$

$+ \quad \dfrac{(x, s) \xrightarrow{a} (x', s')}{(x + y, s) \xrightarrow{a} (x', s')} \qquad \dfrac{(y, s) \xrightarrow{a} (y', s')}{(x + y, s) \xrightarrow{a} (y', s')}$

$\dfrac{(x, s) \xrightarrow{c} (x', s')}{(xy, s) \xrightarrow{c} (x'y, s')} \qquad \dfrac{(x, s) \xrightarrow{\sqrt{}} (x', s') \quad (y, s) \xrightarrow{a} (y', s'')}{(xy, s) \xrightarrow{a} (y', s'')}$

$\| \quad \dfrac{(x, s) \xrightarrow{c} (x', s')}{(x \| y, s) \xrightarrow{c} (x' \| y, s')} \qquad \dfrac{(y, s) \xrightarrow{c} (y', s')}{(x \| y, s) \xrightarrow{c} (x \| y', s')}$

$\dfrac{(x, s) \xrightarrow{a} (x', s') \quad (y, s) \xrightarrow{b} (y', s'')}{(x \| y, s) \xrightarrow{\gamma(a,b)} (x' \| y', s''')} \qquad \begin{array}{l} \text{if } \gamma(a, b) \neq \delta,\ a, b \neq \sqrt{}, \\ \text{and } s''' \in effect(\gamma(a, b), s) \end{array}$

$\dfrac{(x, s) \xrightarrow{\sqrt{}} (x', s') \quad (y, s) \xrightarrow{\sqrt{}} (y', s')}{(x \| y, s) \xrightarrow{\sqrt{}} (x' \| y', s')}$

$\mathbb{L} \quad \dfrac{(x, s) \xrightarrow{c} (x', s')}{(x \mathbb{L} y, s) \xrightarrow{c} (x' \| y, s')}$

$| \quad \dfrac{(x, s) \xrightarrow{a} (x', s') \quad (y, s) \xrightarrow{b} (y', s'')}{(x \mid y, s) \xrightarrow{\gamma(a,b)} (x' \| y', s''')} \qquad \begin{array}{l} \text{if } \gamma(a, b) \neq \delta,\ a, b \neq \sqrt{}, \\ \text{and } s''' \in effect(\gamma(a, b), s) \end{array}$

$\dfrac{(x, s) \xrightarrow{\sqrt{}} (x', s') \quad (y, s) \xrightarrow{\sqrt{}} (y', s')}{(x \mid y, s) \xrightarrow{\sqrt{}} (x' \| y', s')}$

$\partial_H \quad \dfrac{(x, s) \xrightarrow{a} (x', s')}{(\partial_H(x), s) \xrightarrow{a} (\partial_H(x'), s')} \quad$ if $a \notin H \subseteq A$

recursion $\quad \dfrac{(<t_x \mid E>, s) \xrightarrow{a} (y, s')}{(<x \mid E>, s) \xrightarrow{a} (y, s')} \quad$ if $x = t_x \in E$

Table 8.5
Transition rules ($a, b, c \in A_{\sqrt{}}$, $c \neq \sqrt{}$, $H \subseteq A$).

A1	$x + y = y + x$	G1	$\varphi \cdot \neg\varphi = \delta$
A2	$x + (y + z) = (x + y) + z$	G2	$\varphi + \neg\varphi = \epsilon$
A3	$x + x = x$	G3	$\varphi(x + y) = \varphi x + \varphi y$
A4	$(x + y)z = xz + yz$	G4	$a(\varphi x + \neg\varphi y) \subseteq ax + ay$
A5	$(xy)z = x(yz)$	Kl1	$x^* = x \cdot x^* + \epsilon$
A6	$x + \delta = x$	Kl2	$x^* = (x(x + \epsilon))^*$
A7	$\delta x = \delta$	Kl3	$(x + y^*)^* = (x + y)^*$
A8	$\epsilon x = x$	Kl4	$(x \cdot y^*)^* = \epsilon + x(x + y)^*$
A9	$x\epsilon = x$	Kl5	$(x^*(y + \epsilon))^* = (x(y + \epsilon) + y)^*$

Table 8.6
The axioms of BPA_G^* where $\varphi \in G$ and $a \in A$.

9 A Roadmap of Some Two-Dimensional Logics

Vaughan Pratt

9.1 Background

The theme of this chapter is logics with more or less independent disjunction and conjunction. At JELIA'90 I described one such logic, action logic, a single-sorted finitely based equational conservative extension of the equational logic of regular expressions, with the language part of the extension consisting of new operations preimplication $A{\rightarrow}B$ (*had A then B*) and postimplication $B{\leftarrow}A$ (*B if-ever A*) [16]. The organizers of the present conference requested that I talk again on action logic. Although I had nothing new to report on this subject it seemed to me that a walk around the neighborhood of action logic might be of some interest. Action logic being what I called a two-dimensional logic, a natural selection of neighbors might be the two-dimensional logics that arise either in the literature or via a natural juxtaposition of notions.

We first give a somewhat more careful definition of dimension in logic than in [16], then embark on a tour of 2D logics organized around the three basic and more or less independent notions of Boolean algebra, residuation, and star (reflexive transitive closure or ancestral).

9.1.1 Dimension in Logic

I propose to take the dimension of a logic to be the smallest number of operations and constants of the logic sufficient to determine the remaining operations and constants. Before making this more precise let us consider the following examples.

Classical logic specifies Boolean algebras for its models, while intuitionistic logic specifies Heyting algebras. In both cases the four operations of conjunction, disjunction, implication, and negation, and the two constants *true* and *false*, are determined by the underlying poset of the model. That is, given a set X, these operations on X are uniquely determined, if they exist, by a partial order $\leq$ on X. Conjunction $A \cdot B$ and disjunction $A + B$ are respectively meet (greatest lower bound) and join (least upper bound) in the poset. Implication $A{\rightarrow}B$ is the "semi-inverse" of multiplication as defined by $A \cdot B \leq C \Leftrightarrow B \leq A{\rightarrow}C$. The truth values 0 and 1 are respectively the bottom and top of the poset, while negation $\neg A$ or A^- is $A{\rightarrow}0$. While these operations need not exist in every partial ordering of X (e.g. the discrete or empty order of a set with at least two elements), when they do exist they are uniquely determined by the order.

Conversely a set X together with any one of the above three binary operations on X uniquely determine the order on X. Noting that $A \cdot B = A$, $A + B = B$, and $A{\rightarrow}B = 1$ are equivalent in any Boolean or Heyting algebra, we may extract the partial order $A \leq B$

---

[0]This work was supported by the National Science Foundation under grant number CCR-8814921 and a grant from Mitsubishi.

from any one of them. (The presence of the symbol 1 in $A{\rightarrow}B = 1$ should be cause for concern; we address this in the first paragraph of the subsection on explicit definition.)

A model of these logics can be taken to be an *algebra* (a structure all of whose relations are operations) with a single binary operation. In this sense both classical and intuitionistic logic are *one-dimensional*: a single operation on a given set suffices to determine the remaining operations and constants.

An ordered monoid $(X, \leq, ;)$ on the other hand consists of two predicates (if we identify the latter binary operation $A; B$ with the ternary predicate $A; B = C$) neither of which determines the other, even after constraining $(X, \leq)$ to be say a Boolean algebra.

We distinguish these situations formally as follows.

DEFINITION 9.1   Let T be a first-order theory with set S_T of symbols. A subset S' of S_T is called *primitive* when distinct models of T assign distinct interpretations to some symbol of S'.

That is, P is primitive when T together with the interpretations of the symbols of P determines the interpretations of, or *implicitly defines*, the remaining symbols of S_T. Note that "primitive" is primarily an attribute of a set of symbols. Moreover T need not determine a unique such set, or even a unique minimal such set, as the Boolean algebra example shows. By "primitive symbol" we mean only "member of a designated primitive set."

DEFINITION 9.2   The *dimension* of T is the minimum, over all primitive subsets P of S_T, of the cardinality of P.

The term *symbol rank* also suggests itself. I am not aware of any standard terminology for this notion.

A reasonable objection is that this notion of dimension depends on the presentation of the theory, in particular on the choice of symbols. After all, we could surely reorganize the ordered monoid $(X, \leq, ;)$ as a structure (X, R) with one quaternary relation $R(A, B, C, D)$ expressing $A; B \leq C; D$ and be able to recover the predicates $A; B$ and $A \leq B$ by setting suitable variables in $A; B \leq C; D$ to 1, the unit of $A; B$.

This objection notwithstanding, the distinction between one and two dimensional logics is well-defined as an artifact of the presentation as defined above. It is a nice problem whether there is a more abstract natural notion of dimension in which Boolean algebras and ordered monoids retain the dimensions we have assigned them here.

9.1.2   Explicit Definition

Another characterization of the notion of primitive set of symbols of T is that the axioms of T can be written using only primitive symbols. The equivalence of this characterization with the above is an illuminating application of, and one way of viewing, Beth's theorem relating implicit and explicit definability.

A first order structure $\mathcal{M} = (X, \langle R_i \rangle_{i \in I})$ consists of a set X together with a family or indexed set of relations R_i where i ranges over an index set I. (The notion of index may be regarded as formalizing the notion of symbol; indeed we may take I to be the set of symbols of a given theory.) Any subset $I' \subseteq I$ of I determines the I'-reduct $(X, \langle R_i \rangle_{i \in I'})$ of $\mathcal{M}$, consisting of the same set X and the family of just those relations R_i of $\mathcal{M}$ for which $i \in I'$.

DEFINITION 9.3   We say that first order structures $\mathcal{M}$ and $\mathcal{M}'$ are P-equal when they have the same P-reduct.

Except for the matter of symbols, P-equality is stronger than isomorphism, which does not even require the same carriers let alone the same interpretations of the common symbols. And isomorphism in turn is stronger than elementary equivalence, which does not even require the carriers to have the same cardinality.

DEFINITION 9.4   We say that T and T' are P-identical when their models are in a bijective correspondence such that corresponding pairs are P-equal.

Beth's theorem relates explicit and implicit definition. We say that the relation symbol R is *explicitly* defined when it appears only once in some axiomatization of T, and in an axiom of the form $R(x_1, \ldots, x_n) \Leftrightarrow \varphi(x_1, \ldots, x_n)$ where $R(x_1, \ldots, x_n)$ is an atomic formula and $\varphi(x_1, \ldots, x_n)$ is a first-order formula with $x_1, \ldots, x_n$ as its only free variables and containing only primitive symbols. For example, taking disjunction as primitive, we may define conjunction explicitly in terms of disjunction as $AB = C \Leftrightarrow (A + C = A\&B + C = B\&(A + D = A\&B + D = B \Rightarrow D + C = D))$. A less transparent example making essential use of existential quantification takes implication $A \to B$ as primitive and defines $A \leq B \Leftrightarrow \exists T[\forall C[C \to T = T]\&A \to B = T]$. (A first order axiomatization of Boolean algebra in terms solely of implication could begin with this definition along with the assertion that $\leq$ so defined is a partial order.)

Explicit definition is evidently a special case of implicit definition, whence explicitly definable relations are *a fortiori* implicitly definable. Beth's theorem asserts the nontrivial converse: *implicitly definable relations are explicitly definable* [2].

THEOREM 9.1   Let T be a first-order theory and let P be a primitive set of symbols of T. Then there exists a first-order theory T' P-identical to T.

Proof   Replace every atomic formula whose predicate symbol is not in P by the equivalent formula involving only predicate symbols of P. Such a formula exists by Beth's theorem.   ∎

Hence one can rewrite any first-order theory in terms of its primitives. The rewriting may introduce nested quantifiers serving as substitutes for the missing symbols—Beth's theorem is not a trivial result. Hence it may make the axioms harder to read: a straight-forward use of implication may turn into an obscure assertion involving say conjunction and several occurrences of quantifiers.

9.1.3   The Main 2D Logics

Modal logic can be viewed as a 2D logic, with the modality $\Box A$, necessarily A, as a unary operation independent of the basic logical or implication order. Modal logic originated with Aristotle c.330 BC, making it the earliest 2D logic by a substantial margin.

The next 2D logic to appear, and the first with two conjunctions, is the calculus of binary relations, introduced in primitive form by De Morgan in 1860 [5]. The two dimensions of this calculus are what Peirce [15] subsequently called its logical and relative parts. Conjunction (and dually disjunction) took two forms, logical conjunction as the intersection of relations L and M and relative conjunction as their composition, notated $L; M$ by Peirce but LM by De Morgan.

In the following argument [*op. cit.*] De Morgan makes quite explicit his view of composition as a form of conjunction. (By "compound," "component," "aggregation," "aggregant," and "impossible" De Morgan means here respectively conjunction, conjunct, disjunction, disjunct, and false.)

A mathematician may raise a moment's question as to whether L and M are properly said to be *compounded* in the sense in which X and Y are said to be compounded in the term XY. In the phrase *brother of parent*, are *brother* and *parent* compounded in the same manner as *white* and *ball* in the term *white ball*. I hold the affirmative, so far as concerns the distinction between *composition* and *aggregation*: not denying the essential distinction between *relation* and *attribute*. According to the conceptions by which *man* and *brute* are aggregated in *animal*, while *animal* and *reason* are compounded in *man*, one primary feature of the distinction is that an impossible component puts the compound out of existence, an impossible aggregant does not put the aggregate out of existence. In this particular the compound relation 'L of M' classes with the compound term 'both X and Y.'

The last two sentences make the argument that composition is more like conjunction than disjunction on the ground that *false* is an annihilator for composition and conjunction ($L; 0 = X0 = 0$) but an identity for disjunction ($X + 0 = X$). (One is tempted to add that composition and conjunction both distribute over binary disjunction; however zeroary disjunction suffices for the point.)

Peirce [15] subsequently developed the main details of the calculus of binary relations, identifying the full set of its operations, giving it its modern notation, and writing extensively about it. Schröder [18] took up where Peirce left off, producing a substantial book on the calculus of binary relations.

Aristotle's modal logic was revived in 1918 by Lewis. Fourteen years later Lewis and Langford [11] defined strict or necessary implication $A \Rightarrow B$ as $\neg \Diamond (A \wedge \neg B)$ (which for classical modal logic simplifies to the clearer $\Box(A{\rightarrow}B)$). This made strict implication a second implication independent of the basic logical or *material* implication $A{\rightarrow}B$ definable as $\neg A \wedge B$. They also defined *consistency* $A \circ B$ as $\neg(A \Rightarrow \neg B)$, thus yielding the first 2D logic with two commutative conjunctions, ordinary Boolean conjunction being the other.

Relevance logic [1, 6] is a 2D logic in which entailment provides the second dimension. Entailment is much like strict implication except that its essence need not be concentrated in a unary modality but rather is ordinarily axiomatized in its own right (though in relevance system E one may define $\Box A$ as $(A \Rightarrow A) \Rightarrow A$ to yield an S4 modality [6, p.126]).

Linear logic [7] consists of additive connectives and multiplicative connectives, corresponding respectively to the logical and relative connectives of the calculus of binary relations.

9.1.4 Nonconstructivity

By a nonconstructive logic I will mean a specification of a class of posets with structure, the *models* of the logic. The structure typically consists of operations of conjunction AB, disjunction $A + B$, implication $A{\rightarrow}B$, and negation $\neg A$ or A^-, and constants true 1 and false 0. What makes posets nonconstructive is that the relationship $A \leq B$, expressing $A \vdash B$, is merely either true or false, independently of how the relationship might be proved. Constructive logic expresses this relationship by a set of proofs of $A \vdash B$ each defined abstractly as a morphism from A to B in a category whose objects are propositions. In constructive logic the proof of a theorem φ is an integral part of the meaning of φ. This chapter will confine attention to nonconstructive 2D logics. The natural models of constructive 2D logics are monoidal categories, with the analogue of a residuated lattice being a closed category.

9.2 The Roadmap

The diagram in Figure 9.2 orders by inclusion a number of classes of ordered monoids, with larger classes towards the top. The main feature of the diagram is the cube, which has various appendages. We begin at the top of the cube with the variety **ISR** of idempotent semirings and then independently add A^* (reflexive transitive closure or star), $A\backslash B$ and A/B (right and left residuals), and A^- (Boolean negation) to form the respective "coatoms" **ISRT**, **RES**, and **BSR**. These independent language extensions then combine in all possible ways to yield the remaining vertices of a cube. A number of related classes are then attached to appropriate points around the cube.

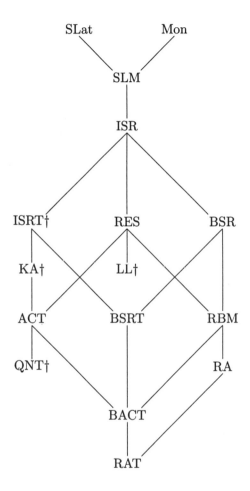

Figure 9.1
The Roadmap (Dagger † denotes a nonvariety)

We now proceed to document this roadmap.

9.2.1   SLat and Mon: Semilattices and Monoids

The essential elements of 2D logic for us will comprise the operations of disjunction $A + B$ and conjunction AB. We require the former to be least upper bound in a poset equipped with a least element. This is equivalent to requiring disjunction to be associative, commutative, and idempotent, and having a unit 0, in which case the relation $A \leq B$ defined as $A + B = B$ can be seen to be a partial order. We call such an algebra $(X, +)$ a *unital semilattice*, with unital specifying the presence of 0. These form the class **SLat**. This class is a variety, that is it can be defined equationally, in fact by finitely many equations, namely the following.

$$A + (B + C) \quad = \quad (A + B) + C \tag{9.1}$$
$$A + B \quad = \quad B + A \tag{9.2}$$
$$A + A \quad = \quad A \tag{9.3}$$
$$A + 0 \quad = \quad A \tag{9.4}$$

In much of what follows we could get by with just a partial order rather than a semilattice, provided we give up varietyhood. However most work done with binary relations and formal languages, the commonest models of what is to come, assumes that finitary joins exist, whence this restriction to semilattices is not unreasonable.

The larger class **OM** of ordered monoids $(X, \leq, ;)$ is the basis for all classes in this chapter. The advantage of **ISR** is that with the join operation the class becomes a variety or equational class.

The class **Mon** consists of monoids $(X, ;)$, for which we require that $A; B$ is associative and has a unit 1 satisfying $A; 1 = A = 1; A$. (We do not need a separate constant symbol for 1 since if such a unit exists it must be unique, via the argument $1 = 11' = 1'$. This class too is a finitely based (finitely axiomatized) variety, with the following equations as its axioms.

$$A(BC) \quad = \quad (AB)C \tag{9.5}$$
$$1A \quad = \quad A \quad = \quad A1 \tag{9.6}$$

9.2.2   SLM and ISR: Semilattice Monoids and Idempotent Semirings

A semilattice monoid is an algebra $(X, +, ;)$ such that $(X, +)$ is a semilattice, $(X, ;)$ is a monoid, and $A; B$ is monotone in A and B with respect to the order defined by $A \leq B$

when $A + B = B$. These form the class **SLM**. The new axiom needed, monotonicity of $A; B$, can be expressed thus.

$$AB \leq (A + A')(B + B') \tag{9.7}$$

By regarding the inequality $a \leq b$ as merely an abbreviation for $a + b = b$ we make this axiom too an equation. These are the axioms for the variety **SLM** of *unital semilattice-ordered monoids*. That is, a model of these equations is a partially ordered monoid, namely a set $(X, \leq, ;)$ where $(X, \leq)$ is a partial order, $(X, ;)$ is a monoid, and AB is monotone in A and B with respect to $\leq$, such that $(X, \leq)$ forms a unital semilattice (has all finite joins including the empty join as bottom). Some authors omit "unital" for convenience, although "semilattice" standardly does not standardly require a bottom.

Distributivity is needed so routinely that we add it right away to form the variety **ISR** of *idempotent semirings*. (But the addition can be postponed while adding star and Boolean negation, which would add an extra square above our cube with **SLM** at the rear corner, which we omit in the interest of keeping down clutter.)

Distributivity is likewise equational. We want finite distributivity; it suffices to require distributivity over binary joins $A + B$ and the empty join 0.

$$A(B + C) = AB + AC \tag{9.8}$$
$$A0 = 0 \tag{9.9}$$
$$(A + B)C = AC + BC \tag{9.10}$$
$$0A = 0 \tag{9.11}$$

Two fundamental instances of idempotent semirings are algebras of binary relations (sets of pairs) and formal languages (sets of strings). In both instances $A + B$ is union, with the empty set as its unit, while AB is either composition or concatenation, with respectively the identity relation I, or the language $\{\varepsilon\}$ having as its one string the empty string, as its unit.

We shall visualize the dimensions as oriented respectively vertically and horizontally. (This will be recognized as in agreement with 2-category usage, where 1-cell composition is horizontal and 2-cell vertical.) It is natural to associate information or static logical strength with the vertical axis and time or dynamic progress with the horizontal.

If we were to add the requirement that AB be meet (greatest lower bound) in the static order the two dimensions would collapse to one and we would have a lattice with least element 0 and greatest element 1.

There are three commonly encountered basic extensions to this logic. All of them are expressible with additional first order axioms, or with equations if we first add appropriate operations. We shall follow the latter approach, leaving the former as an exercise. There are many interesting details about each of these extensions that we have previously written about at length [17, 16] and which we shall therefore not repeat here.

9.2.3   ISRT: Starred Semirings

The most general setting for defining transitive closure A^+ and reflexive transitive closure A^* is an ordered monoid. We say that A is *reflexive* when $1 \leq A$ and *transitive* when $AA \leq A$. Hence we expect A^* to satisfy $1 \leq A^*$ and $A^*A^* \leq A^*$. We also require $A \leq A^*$. These three separate equations can be rolled into one in **SLM**:

$$1 + A^*A^* + A \ \leq \ A^* \tag{9.12}$$

In addition we require A^* to be the *least* element satisfying (9.12), expressible with the following universal Horn formula.

$$1 \leq B \ \& \ BB \leq B \ \& \ A \leq B \ \Rightarrow \ A^* \leq B \tag{9.13}$$

which in **SLM** can also be written

$$1 + BB + A \ \leq \ B \ \Rightarrow \ A^* \ \leq \ B \tag{9.14}$$

This amounts to an induction axiom for star.

We define the class **ISRT** of *starred semirings* to be **ISR** further constrained by the axioms (9.12) and (9.14).

It is clear from the form of (9.14), which requires A^* to be a least element, that $A + B$ and AB remain a primitive set for **ISRT**, which is therefore a 2D logic.

Kozen's class **KA** of *Kleene Algebras* strengthens (9.14) to the following two induction axioms.

$$AB \leq B \ \Rightarrow \ A^*B \leq B \tag{9.15}$$
$$BA \leq B \ \Rightarrow \ BA^* \leq B \tag{9.16}$$

While axiom (9.14) can be proved from these, the converse derivation is not possible in **ISR** but becomes possible in **RES** [16]. These axioms together with those for **ISR** define the class **KA** of *starred semirings*, idempotent semirings with reflexive transitive closure.

In the absence of distributivity, i.e., adding (9.12) and (9.14) to **SLM**, we would have the class **SLMT** of starred semilattice monoids.

9.2.4   RES: Residuated Monoids

The class **RES** of *residuated monoids* is obtained from **ISR** by requiring the existence of left and right *residuals*, a sort of "semi-inverse" of AB. Whereas a true inverse would satisfy $AB = C \Leftrightarrow A = C/B$, we think of a semi-inverse as satisfying $AB \leq C \Leftrightarrow A \leq C/B$. This is "semi-division" on the right yielding the *left residual* (referring to the quantity which "remains" on the left after dividing on the right). Similarly $AB \leq C \Leftrightarrow B \leq A\backslash C$ defines semi-division on the left yielding the right residual, see examples of the use of these constructs in natural language in [16]. A natural alternative notation for $A\backslash C$ is $A{\to}C$, "preimplication", pronounced *had A then C*. Similarly C/B can be written $C{\leftarrow}B$, "postimplication," pronounced *C if-ever B*. It is easily seen that $AB = BA$ implies $A{\to}B = B{\leftarrow}A$, i.e., commutativity of conjunction implies that the two residuals coincide.

We may axiomatize these implications with four universal Horn formulas, shown here in the form of two equivalences.

$$ A \leq C{\leftarrow}B \quad \overset{L}{\Leftrightarrow} \quad AB \leq C \quad \overset{R}{\Leftrightarrow} \quad B \leq A{\to}C \tag{9.17} $$

Another way to define the left residual $C{\leftarrow}B$ is as the join of the set of all A's such that $AB \leq C$. Similarly $A{\to}C$ is the join of the set of all B's such that $AB \leq C$.

In the presence of an operation such as binary join, $A + B$, determining the order, residuation becomes equationally definable. The following equations along with those for $A + B$ and AB suffice to axiomatize the two implications in terms of $A + B$ and AB. Again we are taking $a \leq b$ as an abbreviation for $a + b = b$.

$$
\begin{aligned}
A{\to}B &\leq & A{\to}(B + B') && (9.18) \\
A(A{\to}B) \leq & B &\leq A{\to}AB && (9.19) \\
B{\leftarrow}A &\leq & (B + B'){\leftarrow}A && (9.20) \\
(B{\leftarrow}A)A \leq & B &\leq BA{\leftarrow}A && (9.21)
\end{aligned}
$$

Note that our equational definition explicitly asserts the monotonicity of implication in the numerators B of $A{\to}B$ and $B{\leftarrow}A$, a property that is implicit in the above universal Horn axiomatization (the two equivalences). However antimonotonicity of the denominator A is a consequence of either axiomatization.

We may obtain **RES** from **SLM** instead of from **ISR** using the same equations, since a corollary of being residuated is that conjunction distributes over joins of all cardinalities, including the empty join and infinite joins. (For the "right" way to argue this see [16] following equations (4)-(11).)

Although implication $A{\to}B$ is monotone in its "numerator" B it is antimonotone in its "denominator" A. We may use this fact to obtain a negation operation A^-, defined as $A{\to}0$.

If we now require that AB be meet (a possibility we considered earlier), we obtain exactly the notion of Heyting algebra. The usual notion of propositional intuitionistic logic is defined so as to make its models exactly the Heyting algebras.

LL is Girard's linear logic [7]. Like quantum logic, QL, it is characterized by having an order that is a lattice that is not necessarily distributive, yet having a complement operation satisfying double negation $A^{--} = A$.

9.2.5   BSR: Boolean Semirings

The class **BSR** of *Boolean semirings* consists of those idempotent semirings whose partial order forms a Boolean algebra. (We could as well have defined the larger class of Heyting semirings; we single out the Boolean case out of respect for the very long-standing tradition of defining relation algebras as Boolean rather than Heyting algebras.) To define this class equationally it suffices to add logical negation, A^-, to the language and then to add enough equations to specify (in the case of a Boolean algebra) a complemented distributive lattice. (For a Heyting algebra we would replace the requirement of complementation by that of residuation of the meet $A \cdot B$, defined just as done below for the residuals of AB but with the remark that $A \cdot B$ is commutative whence there is only one residual $A{\to}B$, intuitionistic implication.) Note that this additional organization of the poset affects only the vertical (logical) part and not the horizontal or dynamic part. The reader should be able with some work to supply a suitable axiomatization.

Boolean algebras enjoy a special status as the theory of all possible operations on $\{0, 1\}$, and as the theory of sets under the Boolean set operations of union, intersection, set difference, and complement. Similarly Boolean monoids arise very naturally in 2D logic: the set of all binary relations on a given set forms a Boolean monoid under the Boolean set operations and relation composition. Likewise the set of all formal languages on a given alphabet (set of symbols) forms a Boolean monoid under the Boolean set operations and language concatenation. We discuss Boolean monoids in more detail elsewhere [17].

9.2.6   ACT: Action Algebras

A yet stronger requirement is residuation, which entails distributivity. An *action algebra* [16] is a residuated starred monoid. In the presence of residuation it becomes possible to replace the universal Horn induction axiom (9.14) by the following pair of equations, making the class **ACT** of action algebras a finitely based variety.

$$A^* \ \le \ (A + B)^* \tag{9.22}$$

$$(A{\to}A)^* \; \leq \; A{\to}A \tag{9.23}$$

Axiom (9.23), which I suggested calling *pure induction* [16], originated with Ng and Tarski [14, 13].

We have shown [16] that $A + B$ and AB form a primitive set of operations of **ACT**. In particular **ACT** together with interpretations for $A + B$ and AB determine A^*, along with $A{\to}B$ and $B{\leftarrow}A$. Hence **ACT** is a two-dimensional logic.

To residuation Kozen [10] further adds meet or static conjunction, $A \cdot B$ or $A \wedge B$, without which the matrix-based arguments of [9] cannot be made equational using residuation (though their conclusions continue to hold in the absence of meet). (Meet of course does not increase dimension, being determined by the poset.)

9.2.7   BSRT: Boolean Starred Semirings

The class **BSRT** of Boolean starred semirings is defined in [17] (under the name "normal additive Boolean monoids with star") where it is shown that it is a finitely based variety. Star can be equationally axiomatized in **BSRT** via the following adaptation of Segerberg's induction axiom for propositional dynamic logic [19].

$$A^*B \; \leq \; B + A^*(B^- \cdot AB) \tag{9.24}$$

Again this axiom serves to define A^*, whence **BSRT** remains two-dimensional.

The fact that the induction axiom (9.14) for star can be expressed equationally in either a Boolean or residuated monoid suggests that the main prerequisite for induction is antimonotonicity: while induction wants antimonotonicity it seems not to be too fussy about where it gets it from.

I do not know whether **BSRT** is a subclass of **KA**.

9.2.8   RBM: Residuated Boolean Monoids

A residuated Boolean monoid $(X, \leq, ;)$ is a residuated monoid such that $(X, \leq)$ is a Boolean algebra. Residuation implies distributivity, whence an equally good term for these algebras is residuated Boolean semiring.

RBM is close to but not quite the Jónsson-Tarski [8] class **RA** of *relation algebras*. A relation algebra may be defined as a residuated Boolean monoid equipped with a unary operation $A^{\smile}$ satisfying $A\backslash B = (A^{\smile}; B^-)^-$ and $A/B = (A^-; B^{\smile})^-$. This axiomatization of **RA**, which appeared in this form in [17], is easily seen to define **RA** by comparing it with Chin and Tarski's axiomatization of **RA** [4].

9.2.9 QNT: Quantales

The class **QNT** of *quantales* consists of those semirings whose order is a complete lattice such that conjunction distributes over arbitrary joins (i.e. including infinite joins and the empty join or bottom). Because A^*, $A \backslash B$ and A/B can be defined as joins it follows that **QNT** contains these operations and hence is a subclass of **ACT**. The earliest appearance of the notion of quantale that I am aware of is due to Conway [3], who called them S-algebras. The term "quantale" was introduced more recently by Mulvey [12].

9.2.10 BACT and RAT: Boolean Action Algebras and Starred Relation Algebras

The class **BACT** of Boolean action algebras combines **ACT** with **BSR** by requiring the order of an action algebra to be a Boolean algebra, or equivalently by adding transitive closure to residuated Boolean monoids.

Along the latter lines, Ng and Tarski [14, 13] introduced transitive closure (of the nonreflexive kind) into **RA**. They axiomatized transitive closure with what amounted to axioms (9.12) and (9.14) to define the class **RAT**. They then showed using essentially axioms (9.22) and (9.23) the that **RAT** was a finitely based variety, which is far from obvious if one has not seen (9.23) before.

9.3 Work Needed

Our tour focused on homogeneous or single-sorted 2D logics with the horizontal dimension being defined by "dynamic conjunction" AB. This resulted in the omission of all modal logics, in particular the various dynamic logics, temporal logics, and logics of knowledge. Nor did we attempt to locate the various relevance logics on our roadmap, or the various fragments of linear logic, some of which have attracted considerable attention lately.

We also gave relatively few properties of the logics we did list. We mentioned nothing about decision problems for example, nor about finite model properties or representation theorems. A more complete survey of this area would be a substantial undertaking, not only to collect the extant available results but to identify and answer a great many unsolved problems.

References

[1] A. R. Anderson and N. D. Belnap. *Entailment*, volume 1. Princeton University Press, Princeton, NJ, 1975.

[2] E.W. Beth. On Padoa's method in the theory of definition. *Indagationes Mathematicae*, 15:330–339, 1953.

[3] J.H. Conway. *Regular Algebra and Finite Machines*. Chapman and Hall, London, 1971.

[4]   L.H. Chin and A. Tarski. Distributive and modular laws in the arithmetic of relation algebras. *Univ. Calif. Publ. Math.*, 1:341–384, 1951.

[5]   A. De Morgan. On the syllogism, no. IV, and on the logic of relations. *Trans. Cambridge Phil. Soc.*, 10:331–358, 1860.

[6]   J.M. Dunn. Relevant logic and entailment. In D. Gabbay and F. Guenthner, editors, *Handbook of Philosophical Logic*, volume III, pages 117–224. Reidel, Dordrecht, 1986.

[7]   J.-Y. Girard. Linear logic. *Theoretical Computer Science*, 50:1–102, 1987.

[8]   B. Jónsson and A. Tarski. Representation problems for relation algebras. *Bull. Amer. Math. Soc.*, 54:80,1192, 1948.

[9]   D. Kozen. A completeness theorem for Kleene algebras and the algebra of regular events. In *Proc. 6th Annual IEEE Symp. on Logic in Computer Science*, pages 214–225, Amsterdam, July 1991.

[10]   D. Kozen. On action algebras. This volume.

[11]   C.I. Lewis and C.H. Langford. *Symbolic Logic*. The Century Company, 1932. 2nd ed. 1959, Dover Publications, Inc.

[12]   C.J. Mulvey. In *Second Topology Conference*, Rendiconti del Circolo Matematico di Palermo, ser.2, supplement no. 12, pages 94–104, 1986.

[13]   K.C. Ng. *Relation Algebras with Transitive Closure*. PhD thesis, University of California, Berkeley, 1984. 157+iv pp.

[14]   K.C. Ng and A. Tarski. Relation algebras with transitive closure, Abstract 742-02-09. *Notices Amer. Math. Soc.*, 24:A29–A30, 1977.

[15]   C.S. Peirce. Description of a notation for the logic of relatives, resulting from an amplification of the conceptions of Boole's calculus of logic. In *Collected Papers of Charles Sanders Peirce. III. Exact Logic*. Harvard University Press, 1933.

[16]   V.R. Pratt. Action logic and pure induction. In J. van Eijck, editor, *Logics in AI: European Workshop JELIA '90, LNCS 478*, pages 97–120, Amsterdam, NL, September 1990. Springer-Verlag.

[17]   V.R. Pratt. Dynamic algebras as a well-behaved fragment of relation algebras. In *Algebraic Logic and Universal Algebra in Computer Science, LNCS 425*, Ames, Iowa, June 1988, 1990. Springer-Verlag.

[18]   E. Schröder. *Vorlesungen über die Algebra der Logik (Exakte Logik). Dritter Band: Algebra und Logik der Relative*. B.G. Teubner, Leipzig, 1895.

[19]   K. Segerberg. A completeness theorem in the modal logic of programs. *Notices of the AMS*, 24(6):A-552, October 1977.

10 Some New Landmarks on the Roadmap of Two Dimensional Logics

H. Andreka, I. Nemeti, I. Sain

10.1 The Basic Logics

In this chapter we study decidability and semantical completeness of the logics on Pratt's roadmap of 2D logics (see the previous chapter of this book). Pratt provides a systematic overview of a large class of two dimensional (2D) logics, their interconnections etc. At the end of his contribution, under the title "Work to be done", Pratt mentions two kinds of problems left open. These are (i) which of the introduced logics are *decidable*, and (ii) the so called representation problem for the introduced algebras, which is equivalent with asking whether the introduced logics admit nice, relational *semantics* for which they would be strongly complete.

We address both of these questions here. Question (ii) will be discussed relatively briefly at the end of the paper, for lack of space; but references will be supplied to help the reader in recovering the details.

We begin with recalling concisely what we need from the previous Chapter, and then we start studying question (i). We will try to follow Pratt's notation and conventions as far as we can. Pratt introduces his logics in algebraic form, but the purely logical counterpartshould be easy to find, since the logic to algebra correspondence he uses is described in detail elsewhere, e.g. in [3].

Pratt's starting point is the class *OM* of partially ordered monoids (or ordered monoids, for short). An *OM* is a structure $\mathfrak{A} = \langle A, \leq, ; \rangle$ such that $\langle A, \leq \rangle$ is a partial ordering with smallest element 0, and $\langle A, ; \rangle$ is a *semigroup* with a neutral element 1. The two key ingredients " $\leq$ " and "; " of *OM*s are the source of two dimensionality in Pratt's logics as follows. He associates logical connectives, i.e. operations to " $\leq$ ". Such operations are disjunction " $\vee$ ", negation " $\neg$ ", implication etc. All the operations "associable" to " $\leq$ " form one dimension, which Pratt calls the *static* dimension. He also associates operations (or logical connectives) to the semigroup structure $\langle A, ; \rangle$. These operations, associated to "; " are called *dynamic* ones. They constitute the dynamic dimension. The adjective "dynamic" is justified by the fact that the dynamic aspects of dynamic logic ([12]), of Arrow Logic (cf. e.g. Van Benthem's chapter in this book, Marx-Mikulás-Németi-Sain [10], the proceedings of the Arrow Logic Day at the Logic at Work conference Amsterdam, December 1992, Blackburn-Venema[5]), and of Pratt's Action logic are all derived from the semigroup structure "; ".

In [13] a logic is defined by defining an equational class (variety) of algebras. Valid formulas of the logic correspond to equations valid in the corresponding class of algebras (cf. [3] for detailed justification). A variety is called *decidable* iff its equational theory is decidable. (This is equivalent with decidability of the corresponding logic.)

Let us turn to recalling Pratt's logics (varieties). First we move along the static ($\leq$-

based) dimension.

(1) The first variety is obtained by associating a disjunction "$\vee$" operation to " $\leq$ " (and postulating the usual axioms on " $\vee$ "). This variety *is decidable* (cf. e.g. [6]). To avoid triviality, besides introducing static and dynamic operations (with axioms etc) we need to postulate some interaction between the static and dynamic "worlds". This interaction is provided by postulating that ";" *distributes over* "$\vee$". From now on, this distributivity axiom is assumed unless otherwise specified. (So in defining Pratt's various varieties, we will *not* state this axiom.)

(2) *Idempotent Semirings (ISR's)* are defined just the above way (an *ISR* is an algebra $\langle A, \vee, ; \rangle$ with "$\vee$" a semilattice with 0 and ";" a monoid distributing over "$\vee$"). Moving in the static direction further, if we add *conjunction* " $\wedge$ ", we obtain:

(3) *Distributive Lattice ordered Monoids*, i.e. algebras $\langle A, \vee, \wedge, ; \rangle$ where "$\vee, \wedge$" is a distributive lattice with endpoints and ";" is a monoid. (As promised, we did not write that ";" distributes over " $\vee$ ", but we do require it).

(4) *Boolean Semirings (BSR's)* are algebras $\langle A, \vee, \wedge, \neg, ; \rangle$, where $\langle A, \vee, \wedge, \neg \rangle$ is a Boolean algebra and the rest is as in *ISR*'s.

With this we have exhausted the static direction. Of course, there are important subtleties here which we ignored, e.g. to *ISR*'s we could have added symmetric difference $x \oplus y = (x \wedge \neg y) \vee (\neg x \wedge y)$ making it only a tiny little bit stronger. We will return to this point later. Let us move now in the dynamic direction.

(5) *Residuated ISR's (RES's)* are obtained by adding the residuals "\" and "/" of ";" to *ISR*'s. (Of course, we add the axioms defining the residuals, too.) So *RES*'s are algebras like $\langle A, \vee, ; , \backslash, / \rangle$.

(6) *Action Algebras (ACT)* are obtained from *RES* by adding the iteration " $*$ " of ";". Because of the semantic intuition, " $*$ " is also called transitive closure. Now, we can repeat the same strengthening(s) of the dynamic direction (or dimension), starting out from *BSR* instead of *ISR*, as follows.

(7) *RBM* is obtained from *BSR* by adding the residuals exactly as in (5) above.

(8) *Relation Algebras (RA's)* are obtained from *RBM* by postulating new axioms for the derived operation $x^{\smile} = \neg(\neg x \backslash 1)$ called "converse". Pratt does not use the fact that converse "$\smile$" is explicitly derivable, and gives a more complicated definition of *RA*, not indicating that *RA* is obtained by adding only new axioms (but not operations) to *RBM*.

To illustrate the kind of axioms added here, $x^{\smile\smile} = x$ is not valid in *RBM*, but it is valid in *RA*. For any reflexive and transitive relation R, the algebra $\mathcal{P}(\mathcal{R})$ consisting of all subrelations of R with the natural set theoretic operations is an *RBM* but *not* an *RA* (assuming R is not symmetric). Similar counterexamples can be constructed from any (small) category. A more important difference between *RBM* and *RA* is that *RA* is a so called discriminator variety, while *RBM* is not. Being a discriminator variety is a very strong property which is responsible for many of the characteristic features of Tarskian Algebraic Logic (e.g. RA's, RAT's, cylindric algebras, Boolean algebras etc). These observations seem to indicate that contrary to what Pratt suggests, RBM is not very close to RA after all.

(9) The varieties $BSRT$, $BACT$, and RAT are obtained from BSR, RBM, and RA respectively by adding transitive closure "$*$" as in item (6) above.

The above are the main varieties discussed in Pratt's chapter. Later, we will return to the few further classes mentioned therein.

10.2   The Basic Decidability Results

Let us turn to the results. First we state the theorems, and later we return to their proofs.

THEOREM 10.1 ISR as well as the variety of Distributive Lattice ordered Monoids are *decidable*.

Very probably, the proof method of 10.1 extends to the variety SLM, but we did not check the details.

THEOREM 10.2 OM is decidable. Moreover, all the varieties discussed above become decidable if we omit the interaction axiom between ";" and "$\vee$". E.g. the variety consisting of algebras $\langle A, \vee, \neg, ; \rangle$ with "$\vee, \neg$" a Boolean algebra and ";" a monoid (but no interaction axiom required between the two), is decidable.

A logic or a variety V is *hereditarily undecidable* if it is undecidable and adding new axioms and operations to V leave it undecidable, assuming that these additions leave V nontrivial in a natural sense made precise in [7].

THEOREM 10.3   All the logics extending BSR (Boolean semirings) are undecidable. In particular, BSR, RBM, RA, $BSRT$, $BACT$, RAT as well as Boolean Monoids introduced in §2.5 of Pratt's chapter are all hereditarily undecidable.

Let us have a quick look at our landscape of 2D logics, asking where the borderline between decidable and undecidable is. Undecidability seems to be a consequence of availability of too many of the Booleans (i.e. static operations). Namely, if (besides ";") we have "$\vee$" only (ISR), then it is decidable, if we add "$\wedge$", it remains decidable, but if we add all the Booleans (BSR), then it becomes undecidable. This seems to suggest that adding negation "$\neg$" is responsible for undecidability. Below we will see that this is far from being the case. We will see that starting out from " $\vee$ " (i.e. ISR) and (instead of " $\wedge$ ") adding a little bit of either symmetric difference, or dual implication $x - y = \neg(x \to y)$ causes hereditary undecidability already.

THEOREM 10.4   Consider the decidable variety ISR. Add to the (static) operations Boolean symmetric difference $\oplus$, where $x \oplus y = \neg(x \leftrightarrow y)$ (but of course we do not have " $\leftrightarrow$ " or "$\neg$" in ISR). Add the usual (Boolean) axioms describing $x \oplus y$. Denote the so obtained variety by ISR^+. Then, ISR^+ is undecidable, and all extensions of ISR^+ by axioms and/or operations are undecidable.

We would like to point out that all the undecidability results stated so far follow from 10.4. To prove 10.4 we do not need the full power of symmetric difference. Undecidability follows from much weaker assumptions already. E.g. it is sufficient to use instead of $\oplus$ dual intuitionistic implication. So if $\circ\!\!\to$ denotes intuitionistic implication, then the operation $\neg(x\circ\!\!\to y)$ is sufficient (instead of $x - y$ or $x \oplus y$). For the stronger general results see [7].

The above results take care of all the varieties discussed by Pratt, except for RES and ACT. We conjecture that the methods in the references of this paper are applicable.

One of the *nonvarieties* mentioned in [13] is Linear Logic and it is known to be undecidable [15]. Another one is quantales (QNT). It is not quite clear in what sense decision problems make sense in connection with QNT, but the way QNT is used in §2.9 of Pratt's chapter seems to justify the following argument. Since QNT's are complete lattices, $x\circ\!\!-y = sup\{z : z \leq x$ and $z \wedge y = 0\}$ is expressible in QNT. (The operation "$\circ\!\!-$" is the dual of intuitionistic implication.) We already quoted from [7] the theorem implying that Theorem 10.4 above remains true if we use "$\circ\!\!-$" instead of $\oplus$. But then, this stronger form of Theorem 10.4 yields undecidability of QNT.

There are only two nonvarieties left, but at this point it seems appropriate to formulate some doubt about studying decidability of nonvarieties, as follows. The nonvarieties are not definable by equations, moreover, the ones mentioned by Pratt all need at least quasi-equations (i.e. universal Horn formulas) for defining them. (Indeed, e.g. items (15), (16) in Pratt's list of axioms defining KA are quasi-equations.) So decidability of what should we study? The equational theory which defines a different class and not the one being studied, or its quasi-equational theory which does define the class being studied (in most of the cases). The reasonable choice seems to be to investigate the quasi-equational theories of the nonvarieties, since these are the ones that can be relevant. But, the quasi-equational theories of all the nonvarieties in [13] are *undecidable* (because the quasi-equational theory of semigroups is already undecidable).

With this, the decidability issues of the roadmap seem to be covered. We did not mention $SLat$ and Mon because it is well known that they are decidable. Two problems are left open, namely decidability of the varieties RES and ACT.

The conjecture (and outline of proof) below is in contrast with Theorem 10.1, which said that Distributive Lattice ordered Monoids are decidable. Apparently, if we add "\" to them, then this property goes away.

Conjecture If we add Boolean (static) conjunction $\wedge$ to RES, then it becomes undecidable. The same applies to ACT.

Outline of proof Actually, we do not need the full power of conjunction, but only the operation $dom(x) = 1 \wedge (x \circ 1)$.

We can define a strong kind of complementation $\circ\neg$ as follows: Let $\circ\neg x = (1 \wedge (0 \backslash x)) \circ 1$. Then $\circ\neg$ in cylindric algebraic notation is $\circ\neg x = -c_1 x$. Let $x\circ\!\!-\!\!\circ y = \circ\neg(\circ\neg x \vee y)$. In cylindric algebraic language, $x\circ\!\!-\!\!\circ y = -c_1(-c_1 x \vee y) = c_1 x \wedge -c_1 y)$. Let us try to adapt the first theorem in [7], by putting $\circ\!\!-\!\!\circ$ in place of " $*$ ". Then all the conditions of that theorem are satisfied, except for 2(iii). However, the weaker form 10.1 is satisfied.

$$x \leq (x\circ\!\!-\!\!\circ y) \cup \circ\neg\circ\neg x \tag{10.1}$$

One can modify the proof in [7] in such a way that only (10.1) is needed. E.g. replace $c(x)$ everywhere by $c(\circ\neg\circ\neg x)$. This modified version of the proof in [7] proves the present conjecture. ∎

Remark If instead of conjunction "$\wedge$" we add only converse "$\smile$" to RES (or ACT), then the above outline of proof goes through with some modification (taken from [7] and [14]). E.g. we have to relativize with $T^{\smile\smile}$. Hence $RES + $ "$\smile$" seems to be hereditarily undecidable, too.

Problem Can we eliminate " $\wedge$ " or "$\smile$" from the above proof idea?

10.3 The Proofs and Further Results

When refining the landscape of 2D logics, it is natural to ask what happens if we require ";" to be commutative (or add some similar axioms). These kinds of questions are investigated in [14]. Decidability of subvarieties of RA is studied and surveyed in [2]. The undecidability results in the present chapter all follow from [7]. The decidability results are available from Andréka (see also [6]).

10.4   Representability in Terms of Relations

For lack of space, here we give hints only. For these kinds of 2D algebras there are
two kinds of relational representation in the literature. These are the *square universe*
one and the relativized one, see e.g. Németi[11] for intuitive explanation (but see also
Blackburn-Venema[5], and Marx c.s. [10]). Take BSR as an example. Both approaches
want to represent the elements of a BSR as binary relations. Where they differ is in
the representation of the top element "T" of the Boolean algebra. The square approach
requires T to be represented as a Cartesian square, i.e. to be of the form $T = U \times U$
for some U (whence the name). The relativized approach permits T to be some other
relation, but usually makes some conditions, like requiring T to be reflexive or symmetric.
The result of an operation may not be a subset of T, therefore the relativized approach
intersects the result of any operation with T (this is how complementation is handled
already in Boolean algebras).

The results can be summarized as follows. Square representability fails for practically
all the discussed classes (with the trivial counterexamples of $SLat$ and Mon). See [1],
[11], but relativized representability tends to be available to all of the classes on the
roadmap. See [8], [9], [10], [11].

Acknowledgements

Research supported by the Hungarian National Research Fund OTKA 1911, 2258, T7255.
We are grateful to Ágnes Kurucz and András Simon without whose research work the
present paper could not exist. Actually, much of the results reported here are consequences of ones obtained jointly with them, cf. e.g. [7],[14].

References

[1]   H. Andréka, *Complexity of equations valid in algebras of relations*, Hungar. Acad. Sci. Budapest, 1992.

[2]   H. Andréka, S. Givant and I. Németi. Decision problems for equational theories of relation algebras *Manuscript*, 1992/93. An extended abstract is to appear in Bull. Section of Logic, Poland.

[3]   H. Andréka, I. Németi, I. Sain and Á. Kurucz. On the methodology of applying algebraic logic to logic. *Preprint*, Math. Inst. Hungar. Acad. Sci., 1993. Also available as course material of the LLI Sumerschool 1993 Lisbon. Shortened version to appear in the Proc. of AMAST 1993, Twente, The Netherlands (Springer-Verlag).

[4]   J. van Benthem, A note on dynamic arrow logic. *This volume.*

[5]   P. Blackburn and Y. Venema. Dynamic squares. *Preprint*, Dept. Philos. Utrecht Univ. Logic Group Preprint Series No. 92, 1993.

[6]   V. Gyuris, Decidability of semigroups with a Boolean ordering (and non-distributive modal logics), *Preprint*, Math. Inst. Budapest, 1992.

[7]   Á. Kurucz, I. Németi, I. Sain and A. Simon. Undecidable varieties of semilattice-ordered semigroups, of Boolean algebras with operators, and logics extending Lambek calculus. *Preprint*, Math. Inst. Hungar. Acad. Sci. Budapest, 1993. (To appear in the Bulletin of IGPL.)

[8]   M. Marx. Dynamic arrow logic with pairs. *Proc. Logic at Work*, December 1992. CCSOM, University of Amsterdam, Amsterdam. Also CCSOM Working paper No. 92-71.

[9]   M. Marx, Axiomatizing and deciding relativized relation algebras. *Report* CCSOM, University of Amsterdam, No. 93-87, Amsterdam, 1993.

[10]   M. Marx, Sz. Mikulás, I. Németi and I. Sain. Investigations in arrow logic. *Proc. Logic at Work*, December 1992, CCSOM, University of Amsterdam, Amsterdam.

[11]   I. Németi, Algebraizations of quantifier logics, an introductory overview. *Version 11.* Available as (1) Math. Inst. preprint, Budapest, (2) Univ. of Amsterdam, CCSOM Report No. 91-67, (3) Radically shorter version appeared in Studia Logica L,3-4 (1991), 485-569.

[12]   V.R. Pratt. Dynamic algebras. *Studia Logica*, 50, 3-4, 1991, 571-605.

[13]   V.R. Pratt. A roadmap of some two-dimensional logics. *This volume.*

[14]   A. Simon, I. Sain and I. Németi. Some classes of residuated Boolean algebras with operators, their equational undecidability. *Preprint*, Math. Inst. Budapest, 1993.

[15]   A.S. Troelstra. *Lectures on Linear Logic*, CSLI Lecture Notes No. 29, Stanford, 1992.

11 Meeting Some Neighbours

Maarten de Rijke

11.1 Introduction

Over the past several years the computer science community seems to have lost interest in dynamic logic and related systems somewhat. In the philosophical community, on the other hand, more and more people have felt a need for systems in which changes and processes can be modelled. This has lead to the birth of quite a number of systems blessed with the predicate 'dynamic'.

In this chapter one such system, called DML, is taken as a starting point, and its connections with alternative dynamic proposals are examined. Specifically, a revision operator is defined in DML which can be shown to satisfy most of the postulates such operators are currently supposed to satisfy. Further links are established with terminological logic, Veltman's update semantics, and preferential reasoning. Technical results pertaining strictly to the dynamic modal system of this chapter are given in a companion paper.

The purpose of this chapter is to discuss links between a recent proposal for reasoning about the dynamics of information, called *dynamic modal logic* or DML, and other such proposals, as well as connections with some other formalisms in philosophical logic, cognitive science and AI. The key phrases common to most of the systems that come up in this note are (minimal) change and reasoning about information.

As many dynamic-like formalisms have been proposed over the last few years, the danger that several researchers might be re-inventing the wheel is not entirely fictitious. For that reason I think it is important to have occasional comparisons across platforms. As a result of such comparisons results known in one domain may shed light on problems in the other domains, allowing the field at large to benefit. And at a more down to earth level the obvious advantage of such comparisons is that they may serve as partial maps of rapidly changing research areas. Thus, the purpose of this note is to sketch such a partial map by comparing or unifying some related dynamic systems using the DML formalism.

What's commonly considered to be the minimal requirement for a system to be called dynamic, is that it has a notion of state, and a notion of change or transition from one state to another. States and transitions are precisely the basic ingredients of the system DML; in addition it has various systematic connections between those basics. Although DML may at first appear to be a somewhat unorthodox modal system, it can be analyzed using fairly traditional tools from modal logic, yielding results on its expressive power, the hardness of the satisfiability problem for the language, and axiomatic completeness.

The main benefits of using DML as a guide-line for linking a number of dynamic proposals are the fact that many dynamic proposals are, so to say, de-mystified by being embedded in a system itself comprising of two well-known components (Boolean algebra and relational algebra); the embedding of such proposals into (a fragment of) DML suggests natural additions to, and generalizations of, these proposals. Moreover, the work presented here shows how fairly orthodox dynamic proposals like DML can be used fruitfully far beyond their traditional boundaries.

In 11.2 I describe the basics concerning DML, including two ways of dealing with the states of DML models: one can either take the usual view as states as objects devoid of any structure, or one can endow them with an internal structure and logic of their own.

After that I move on to two connections between DML equipped with 'structureless' states and other systems. In 11.3 an example from cognitive science and AI is considered when I model certain postulates for theory change inside DML. In 11.4, a link is established between DML and terminological logic and knowledge representation. I obtain an exact match between DML and a KL-ONE dialect, called the Brink and Schmidt language, plus an axiomatization of the representable algebras underlying this language.

In 11.5 and 11.6 the states of our DML-models will be equipped with structure. This is needed in 11.5 to link DML to a system of update semantics from the philosophical logic tradition proposed by Frank Veltman, while 11.6 contains some suggestions on how one would have to go about dealing with preferences and other more complex systems in DML.

Section 11.7 rounds off the chapter with some conclusions and questions.

11.2 DML: A Quick Review

11.2.1 Basics

The system of dynamic modal logic DML figuring in this note first appeared in what's more or less its present form in (Van Benthem [4]), but parts of it can be traced back to Van Benthem[2]. The original application of the system was reasoning about the knowledge of a single agent, and the "epistemic moves" this agent makes in some cognitive space to acquire new knowledge. Thus, in DML provisions have been made to talk about transitions that represent the acquisition of new knowledge, and about transitions representing the loss or giving up of knowledge. Moreover, these transitions may be structured in a variety of ways. To sum up, the DML-language has Boolean ingredients to reason about the static aspects of the agents knowledge, and relational ingredients to reason about the dynamic aspects thereof. In addition there are systematic connections between the two realms, as depicted in Figure 11.1.

After some cleaning up had been performed, a stable version of the language was given in (De Rijke [23]). Here it is:

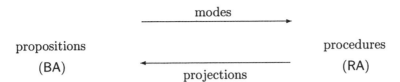

Figure 11.1
DML, the basic ingredients

Atomic formulas: $p \in \Phi$,
Formulas: $\varphi \in Form(\Phi)$,
Procedures: $\alpha \in Proc(\Phi)$.

$\varphi ::= p \mid \bot \mid \top \mid \varphi_1 \rightarrow \varphi_2 \mid \mathsf{do}(\alpha) \mid \mathsf{ra}(\alpha) \mid \mathsf{fix}(\alpha)$,
$\alpha ::= \mathsf{exp}(\varphi) \mid \mathsf{con}(\varphi) \mid \alpha_1 \cap \alpha_2 \mid \alpha_1 ; \alpha_2 \mid -\alpha \mid \alpha^{\smallsmile} \mid \varphi?$.
I will refer to elements of $Form(\Phi) \cup Proc(\Phi)$ as *expressions*.

The intended interpretation of the above connectives and mappings is the following. A formula $\mathsf{do}(\alpha)$ ($\mathsf{ra}(\alpha)$) is true at a state x iff x is in the domain (range) of α, and $\mathsf{fix}(\alpha)$ is true at x if x is a fixed point of α. The interpretation of $\mathsf{exp}(\varphi)$ (read: expand with φ) in a model $\mathfrak{M}$ is the set of all moves along the "informational ordering" in $\mathfrak{M}$ that take you to a state where φ holds; the interpretation of $\mathsf{con}(\varphi)$ (read: contract with φ) consists of all moves *backwards* along the ordering to states where φ *fails*; φ? is the "test-for-φ" relation, while the intended interpretation of the operators left unexplained should be clear.

The models for this language are structures of the form $\mathfrak{M} = (W, \sqsubseteq, \llbracket \cdot \rrbracket, V)$, where $\sqsubseteq \subseteq W^2$ is a transitive and reflexive relation (the informational ordering), $\llbracket \cdot \rrbracket : Proc(\Phi) \rightarrow 2^{W \times W}$, and $V : \Phi \rightarrow 2^W$. The interpretation of the modes is:

$\mathfrak{M}, x \models \mathsf{do}(\alpha)$ iff $\exists y \, ((x, y) \in \llbracket \alpha \rrbracket)$,
$\mathfrak{M}, x \models \mathsf{ra}(\alpha)$ iff $\exists y \, ((y, x) \in \llbracket \alpha \rrbracket)$,
$\mathfrak{M}, x \models \mathsf{fix}(\alpha)$ iff $(x, x) \in \llbracket \alpha \rrbracket$,

while the relational part is interpreted using the mapping $\llbracket \cdot \rrbracket$:

$$
\begin{aligned}
\llbracket \mathsf{exp}(\varphi) \rrbracket &= \lambda xy. \, (x \sqsubseteq y \wedge \mathfrak{M}, y \models \varphi), \\
\llbracket \mathsf{con}(\varphi) \rrbracket &= \lambda xy. \, (x \sqsupseteq y \wedge \mathfrak{M}, y \not\models \varphi), \\
\llbracket \alpha \cap \beta \rrbracket &= \llbracket \alpha \rrbracket \cap \llbracket \beta \rrbracket, \\
\llbracket \alpha ; \beta \rrbracket &= \llbracket \alpha \rrbracket ; \llbracket \beta \rrbracket, \\
\llbracket -\alpha \rrbracket &= -\llbracket \alpha \rrbracket, \\
\llbracket \alpha^{\smallsmile} \rrbracket &= \{ (x, y) : (y, x) \in \llbracket \alpha \rrbracket \},
\end{aligned}
$$

$$[\![\varphi?]\!] \quad = \quad \{\, (x,x) : \mathfrak{M}, x \models \varphi \,\}.$$

Obviously, ra and fix are definable using the other operators, however, for conceptual and notational convenience they will be part of the official definition of the language. Further examples of operators definable in terms of the others will be given below.

I will refer to this language as the DML-language, and in more formal parts of this chapter also as $DML(\sqsubseteq, \Phi)$, where Φ is the set of proposition letters. A natural extension is obtained by considering multiple basic relations $\{\sqsubseteq_i\}_{i \in I}$ instead of the single relation $\sqsubseteq$; I will write $DML(\{\sqsubseteq_i\}_{i \in I}, \Phi)$ for the language thus extended. (In this extended language the expansion and contraction operators will be indexed with the relations they are based upon, viz. $\exp(\varphi)_i$ and $\mathsf{con}(\varphi)_i$.)

In its formulation in Van Benthem [4] the DML-language also contained *minimal* versions $\mu\text{-}\exp(\cdot)$ and $\mu\text{-}\mathsf{con}(\cdot)$ of the expansion and contraction operators, respectively, where $[\![\mu\text{-}\exp(\varphi)]\!] =$

$$\lambda xy.\, \Big((x,y) \in [\![\exp(\varphi)]\!] \wedge \neg \exists z\, (x \sqsubseteq z \sqsubset y \wedge (x,z) \in [\![\exp(\varphi)]\!]) \Big),$$

and likewise for $\mu\text{-}\mathsf{con}(\varphi)$. However, there is no need to add them explicitly to the language, as both are definable:

$$[\![\mu\text{-}\exp(\varphi)]\!] = [\![\exp(\varphi) \cap -(\exp(\varphi); (\exp(\top) \cap -(\top?)))]\!],$$

and similarly for $\mu\text{-}\mathsf{con}(\varphi)$.

11.2.2 Some Results

Let me mention some of the work that has been done on DML. De Rijke [23] gives an explicit axiomatization of validity in DML, comprising of 36 axioms, and 4 derivation rules (including a so-called 'unorthodox' Gabbay-style irreflexivity rule). For future reference let me record this result:

THEOREM 11.1 There exists a complete, finitary axiomatization of validity in the language $DML(\{\sqsubseteq_i\}_{i \in I}, \Phi)$.

De Rijke [23] uses a difference operator $\mathbf{D}$ ('truth at a different state') to characterize some of the modes and projections, for example

$$p \wedge \neg\mathbf{D}p \rightarrow \Big(\mathsf{fix}(\alpha \cap \beta) \leftrightarrow \mathsf{do}(\alpha \cap \beta; p?) \Big)$$

is an axiom in his axiomatization governing the interaction of fix and $\cap$.

The same paper also establishes the undecidability of satisfiability in DML. In addition it gives a number of subsystems and extensions of DML whose satisfiability problems are decidable; in particular, deleting (; and) or just − yields decidable fragments again, as does restricting the class of models to those based on trees. Furthermore, exact descriptions, both syntactic, and semantic by means of appropriate bisimulations, are given for the first-order counterpart of DML.

11.2.3 Some Connections

There are obvious connections between DML and *propositional dynamic logic* (PDL, cf. Harel [18]). The 'old diamonds' $\langle \alpha \rangle$ from PDL can be simulated in DML by putting $\langle \alpha \rangle \varphi := \mathsf{do}(\alpha; \varphi?)$. And likewise, the expansion and contraction operators are definable in a particular mutation of PDL where taking converses of program relations is allowed and a name for the informational ordering is available: $[\![\mathsf{exp}(\varphi)]\!] = [\![\sqsubseteq; \varphi?]\!]$ and $[\![\mathsf{con}(\varphi)]\!] = [\![\sqsubseteq \; ; \neg\varphi?]\!]$. The operator $\mathsf{do}(\alpha)$ can be simulated in standard PDL by $\langle \alpha \rangle \top$. An obvious difference between DML and PDL is that (at least in it's more traditional mutations) PDL only has the *regular* program operations $\cup, ;$ and *, while DML has the full relational repertoire $\cup, -,$ and $;$, but not the Kleene star. Another difference is not a technical one, but one in emphasis; whereas in PDL the Boolean part of the language clearly is the primary component of the language, in DML some effort is made to give the relational part the status of a first-class citizen as well by shifting the notation towards one that more clearly reflects the aspects of relations which we usually consider to be important.

A related formalism whose relational apparatus is more alike that of DML is the *Boolean modal logic* (BML) studied by Gargov and Passy [14]. This system has atomic relations $\rho_1, \rho_2, \ldots$, a constant for the Cartesian product $W \times W$ of the underlying domain W, and relation-forming operators $\cap, \cup$ and $-$. Relations are referred to within the BML-language by means of the PDL-like diamonds $\langle \alpha \rangle$. Since BML does *not* allow either ; or as operators on relations, it is a strict subsystem of $DML(\{ \rho_1, \rho_2, \ldots \}, \Phi)$.

Further connections between DML and related work have been given in (Van Benthem [4]). These include links with Hoare Logic, and with various styles of non-standard inference.

11.2.4 Adding Structure

Usually no assumptions are made on the nature of the states of modal models. But for some applications of modal or temporal logics it may be necessary to be more specific about their nature. (Cf. (Gabbay, Hodkinson and Reynolds [11]) for a whole array of examples.) In such a structured setting models will have the form $\mathfrak{M} = (W_g, \ldots)$, where the *global* components of the model are given by the. . . , while the set W_g is a set of models $\{ \mathfrak{m} \}_{i \in I}$ each of which may have further structure. For instance, they may themselves be of the form $\mathfrak{m} = (W_l, R, V_l)$. Clearly, two languages are involved here: a *global* language which talks about global aspects of the structure, but which does not deal with local aspects, and, secondly, there is *local* language used to reason only about the internal structure of the elements of the model $\mathfrak{M}$. Below, in 11.5 and 11.6, I will equip the states of DML-models with structure in different ways, each with an appropriate local language, but in every case DML will be the global language.[1]

11.3   On Postulates for Theory Change

In this section I will first discuss to which extent Gärdenfors' theory on the dynamics of belief and knowledge can be dealt with in the DML language. After that I will discuss two alternative proposals, and finally I will tie up some loose ends.

11.3.1   The Gärdenfors Postulates

Consider a set of beliefs or a knowledge set T.[2] As our perception of the world as described by T changes, the knowledge set may have to be modified. In the literature on theory change or belief revision a number of such modifications have been identified (cf. (Alchourrón, Gärdenfors and Makinson [1]), and (Katsuno and Mendelzon [20])); these include expansions, contractions and revisions. If we acquire information that does not contradict T, we can simply *expand* our knowledge set with this piece of information. When a sentence φ previously believed becomes questionable and has to be abandonded, we *contract* our knowledge with φ. Somewhat intermediate between expansion and contraction is the operation of *revision*, this is the operation of resolving the conflict that arises when the newly acquired information contradicts our old beliefs. The revision of T by a sentence φ, $T * \varphi$, is often thought of as consisting of first making changes to T, so as to then be able to expand with φ. According to general wisdom on theory change, *as little as possible* of the old theory should be given up in order to accommodate for newly acquired information.

Gärdenfors and others have proposed a set of rationality postulates that the revision operation must satisfy. To formulate these, let a *knowledge set* be a deductively closed set of formulas. Given a knowledge set T and a sentence φ, $T * \varphi$ is the revision of T by φ. $T + \varphi$ ("the expansion of T by φ") is the smallest deductively closed set extending $T \cup \{\varphi\}$.

Basic Gärdenfors postulates for revision

$(*1)$ $T * \varphi$ is a knowledge set.

---

[1]The essential syntactic restriction corresponding to the above global-local distinction is that operators from the global language are *not* allowed to occur inside the scope of operators from the local language. By results of Finger and Gabbay [9], if both the local and the global language have some nice property P, then so does their composition, provided that the above syntactic restriction is met; here P can be a property like enjoying a complete recursive axiomatization, decidability, or the finite model property.
[2]This subsection was inspired by a reading of (Fuhrmann [10]).

($*2$) $\varphi \in T * \varphi.$
($*3$) $T * \varphi \subseteq T + \varphi.$
($*4$)   If $\neg\varphi \notin T$ then $T + \varphi \subseteq T * \varphi.$
($*5$)   If $\bot \in T * \varphi$ then φ is unsatisfiable.
($*6$)   If $\varphi \leftrightarrow \psi$ then $T * \varphi = T * \psi.$

Additional Gärdenfors postulates for revision

($*7$) $T * (\varphi \wedge \psi) \subseteq (T * \varphi) + \psi.$
($*8$)   If $\neg\psi \notin T * \varphi$ then $(T * \varphi) + \psi \subseteq T * (\varphi \wedge \psi).$

For an intuitive explanation of this postulates I refer the reader to (Alchourrón et al [1], Gärdenfors [12]).

To represent the revision operator in DML some choices need to be made. First, we have to agree on some kind of structure in which our theories will be represented, and in which transitions between theories will take place. To keep things simple, and exclude what I consider to be aberrations in this context (like densely ordered sequences of theories), let us assume that our structures are well-founded ones (in addition to being pre-orders, of course).

Next, we have to decide how to represent theories or knowledge sets. The natural option suggested by standard practice in epistemic logic is to do this. Let $\sqsubseteq$ abbreviate $\mathsf{exp}(\top)$, and let $[\sqsubseteq]\varphi$ be short for $\neg\langle\sqsubseteq\rangle\neg\varphi$. Then, I represent theories as sets of the form $w_\square = \{\varphi : \mathfrak{M}, w \models [\sqsubseteq]\varphi\}$, for some w in the model $\mathfrak{M}$. Then "$\varphi \in T$" may be represented as "$[\sqsubseteq]\varphi$," that is, as $\neg\mathsf{do}(\mathsf{exp}(\top); \neg\varphi?)$.

A third choice needs to be made to represent the *expansion* operator $[+\varphi]\psi$ ("ψ belongs to every theory resulting from expanding with φ"). Here I opt for:

$$[+\varphi]\psi := \neg\mathsf{do}(\mu\text{-}\mathsf{exp}([\sqsubseteq]\varphi); \neg[\sqsubseteq]\psi?).$$

So, a formula $[+\varphi]\psi$ is true at some point x if in every 'minimal' $\sqsubseteq$-successor y of x where $[\sqsubseteq]\varphi$ holds (i.e. where φ has been added to the theory), the formula $[\sqsubseteq]\psi$ is true (i.e. ψ is in the theory). Obviously, $[\sqsubseteq]\psi$ may be viewed as the special case of $[+\varphi]\psi$, where one expands with $\varphi = \top$.

Representing the revision operator $[*\varphi]\psi$ ("ψ belongs to every theory resulting from revision by φ") is a slightly more complex matter. Recall that revision of T by φ is explained as removing from T all (and only those) sentences that are inconsistent with φ, and subsequently expanding T by φ.[3] Mimicking the removal from T of the formula that causes the inconsistency with φ by $\mu\text{-}\mathsf{con}([\sqsubseteq]\neg\varphi)$, and the subsequent expansion with φ as before, I end up with the following definition:

---

[3] Isaac Levi has in fact suggested that revisions should be *defined* in terms of such contractions and revisions.

$$[*\varphi]\psi := \neg\mathsf{do}\Big([\mu\text{-con}([\sqsubseteq]\neg\varphi); \mu\text{-exp}([\sqsubseteq]\varphi)]; \neg[\sqsubseteq]\psi?\Big). \ [4]$$

Before actually translating the revision postulates into DML, let me mention a possible point of discussion here. In my approach the expansion and revision operators lack the functional character they have in the Gärdenfors approach. This is due, of course, to the fact that the underlying $\sqsubseteq$-paths to points where "$\varphi \in T$" holds or fails for the first time, need not be uniquely determined. I don't see this as a shortcoming of the way I've set up things. On the contrary, one can view this as an attempt to take the non-deterministic character of everyday expansions and revisions seriously, instead of dismissing it as being "non-logical".

Another source of indeterminism is that, starting from a given node/theory and a formula φ that you want to expand with, you may have to pass several other nodes/theories before ending up at an outcome of the expansion, while a move to contract by φ at this outcome need not take you all the way back to your starting point.[5]

Finally, despite the fact that expansions and revisions may have multiple outcomes in my setup, they need not have a single one, i.e., expansions and revisions need not be defined in every situation.

Given the above points some of the postulates (∗1)–(∗8) are bound to come out invalid when translated into DML. But on the other hand, they also allow for some choices when doing the translation. The statement $\psi \notin T * \varphi$ may be read as "ψ does not belong to *any* theory resulting from revision by φ," or as "for *some* outcome T' of revising T by φ, $\psi \notin T'$." The modal counterparts of these options are

$$\neg\mathsf{do}\Big([\mu\text{-con}([\sqsubseteq]\neg\varphi); \mu\text{-exp}([\sqsubseteq]\varphi)]; [\sqsubseteq]\psi?\Big),$$

or $[\dagger\varphi]\psi$ for short, and $\neg[*\varphi]\psi$, or

$$\mathsf{do}\Big([\mu\text{-con}([\sqsubseteq]\neg\varphi); \mu\text{-exp}([\sqsubseteq]\varphi]; \neg[\sqsubseteq]\psi?\Big),$$

respectively. These subtleties will make some difference for postulate (∗8).

On a similar note, as expansions and revisions need not be defined in every situation, one might consider adding a clause $\neg[+\varphi]\perp$ ($\neg[*\varphi]\perp$) saying "and if expansion (revision) with φ is at all possible" to some of the Gärdenfors postulates. However, for none of the postulates this has any visible effects.

---

[4]This definition is clearly in accordance with the earlier maxim "change as little as possible of the old theory."

[5]In other words: it may be that you need to expand with some formulas $\psi_1, \ldots, \psi_n$ before you can expand with φ. Admittedly, this kind of interference may be undesirable, especially when $\psi_1, \ldots, \psi_n$ and φ are logically independent; on the other hand, this interference might be useful to model various kinds of *non*-logical dependencies between formulas.

(G2) $[*\varphi]\varphi$
(G3) $[*\varphi]\psi \rightarrow [+\varphi]\psi,$
(G4) $\neg[\sqsubseteq]\neg\varphi \wedge [+\varphi]\psi \rightarrow [*\varphi]\psi,$
(G5) $[*\varphi]\bot \rightarrow [*\psi]\neg\varphi,$
(G6) $\varphi \leftrightarrow \psi \,/\, [*\varphi]\chi \leftrightarrow [*\psi]\chi,$
(G7) $[*(\varphi \wedge \psi)]\chi \rightarrow [*\varphi][+\psi]\chi,$
(G8a) $\neg[*\varphi]\neg\psi \wedge [*\varphi][+\psi]\chi \rightarrow [*(\varphi \wedge \psi)]\chi,$
(G8b) $\neg[\dagger\varphi]\psi \wedge [*\varphi][+\psi]\chi \rightarrow [*(\varphi \wedge \psi)]\chi.$

Table 11.1
Translating the Gärdenfors postulates.

Which translations does this give, then? Translating $\chi \in T + \varphi$ as $[+\varphi]\chi$, with $[\sqsubseteq]\chi$ as the limiting case where $T + \varphi$ is in fact T (or $T + \top$), and, likewise, translating $\chi \in T * \varphi$ as $[*\varphi]\psi$, I arrive at Table 11.1, where Gn is the translation of postulate $(*n)$. Observe that there is no schema corresponding to postulate $(*1)$ in Table 11.1; this one seems to resist a direct translation, but its validity is guaranteed given the choices I have made.

Which of the schemata G2–G8b is valid on the well-founded DML-models we are considering here? First, the translation G2 of $(*2)$ comes out valid, as an easy calculation shows. To see that G3 is valid, assume that in some model we have $x \not\models [+\varphi]\psi$. So there is a minimal $\sqsubseteq$-successor y of x with $y \models [\sqsubseteq]\varphi$, $\neg[\sqsubseteq]\psi$. Let us verify that $x \not\models [*\varphi]\psi$. Clearly, $y \models [\sqsubseteq]\varphi$ implies $x \not\models [\sqsubseteq]\neg\varphi$, so $(x,x) \in [\![\mu\text{-con}([\sqsubseteq]\neg\varphi)]\!]$. In addition $(x,y) \in [\![\mu\text{-exp}([\sqsubseteq]\varphi)]\!]$. Hence, as $y \models \neg[\sqsubseteq]\psi$, we must have

$$x \models \mathsf{do}\Big([\mu\text{-con}([\sqsubseteq]\neg\varphi; \mu\text{-exp}([\sqsubseteq]\varphi)]; \neg[\sqsubseteq]\psi?\Big),$$

which is what we were after. Ergo, G3 is valid on all DML-models.

Next comes G4. Suppose that $x \models \neg[\sqsubseteq]\neg\varphi$, $[+\varphi]\psi$, but that $x \not\models [*\varphi]\psi$. We derive a contradiction. By $x \not\models [*\varphi]\psi$ there is a minimal $\sqsubseteq$-predecessor y of x with $y \models \neg[\sqsubseteq]\neg\varphi$. But as $x \models \neg[\sqsubseteq]\neg\varphi$, x itself must be this y. But then, by assumption, $x \models \mathsf{do}(\mu\text{-exp}([\sqsubseteq]\varphi; \neg[\sqsubseteq]\psi?))$, that is: for some minimal $\sqsubseteq$-successor z of x, $z \models [\sqsubseteq]\varphi$, $\neg[\sqsubseteq]\psi$. But by $x \models [+\varphi]\psi$, we must also have $z \models [\sqsubseteq]\psi$, yielding the required contradiction. Hence G4 is valid.

G5 is trivially valid, as its antecedent can never be satisfied. The validity of G6 is also obvious, so let us consider G7. Seeing that it is valid requires a small argument. Assume that in some model we have $x \not\models [*(p \wedge q)]r \rightarrow [*p][+q]r$. Then there are y, z, u such that

1. y is a minimal $\sqsubseteq$-predecessor of x with $y \not\models [\sqsubseteq]\neg p$,

2. z is a minimal $\sqsubseteq$-successor of y with $z \models [\sqsubseteq]p$,

3. u is a minimal $\sqsubseteq$-successor of z with $u \models [\sqsubseteq]q$, $\neg[\sqsubseteq]r$.

To arrive at a contradiction assume that

4. $x \models [*(p \wedge q)]r$.

Then, by (1) and an easy argument, y must be a minimal $\sqsubseteq$-predecessor of x with

5. $y \not\models [\sqsubseteq]\neg(p \wedge q)$.

To arrive at a contradiction, we will show that $u \models [\sqsubseteq]r$ — conflicting with (3). By (4) and (5), if u is a minimal $\sqsubseteq$-successor of y with $u \models [\sqsubseteq](p \wedge q)$, we must have $u \models [\sqsubseteq]r$. If, on the other hand, u is not such a successor, then, as $u \models [\sqsubseteq](p \wedge q)$ by (2) and (3), there must be a v such that

6. v is a minimal $\sqsubseteq$-successor of y with $v \models [\sqsubseteq](p \wedge q)$ and $v \sqsubseteq u$,

because we have assumed our structures to be well-founded. But then, by (4) and (5), $v \models [\sqsubseteq]r$, and by (6), $u \models [\sqsubseteq]r$, and we have reached our contradiction. This implies that G7 is valid.

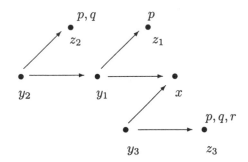

Figure 11.2
Refuting G8a

In G8a the antecedent $\neg\psi \notin T * \varphi$ of (*8) is translated as $\neg[*\varphi]\neg\psi$. The instance $\neg[*p]\neg q \wedge [*p][+q]r \rightarrow [*(p \wedge q)]r$ of G8a is refuted at x in the model depicted in Figure 11.2. To see this, notice first of all that $[*(p \wedge q)]r$ is refuted at x because

$(x, z_2) \in [\![\mu\text{-con}([\sqsubseteq]\neg(p \wedge q)); \mu\text{-exp}([\sqsubseteq](p \wedge q)); \neg[\sqsubseteq]r?]\!]$.

Second, $\neg[*p]\neg q$ holds at x as

$(x, z_3) \in [\![\mu\text{-con}([\sqsubseteq]\neg p); \mu\text{-exp}([\sqsubseteq]p); [\sqsubseteq]q?]\!]$.

Third, $[*p][+q]r$ holds at x because there's only one "revise by p, expand by q" path leading from x, notably (x, z_3), and at the end of that path $[\sqsubseteq]r$ holds. (In particular, (x, z_2) is not a "revise by p, expand by q" path since $(x, y_2) \notin [\![\mu\text{-con}([\sqsubseteq]\neg p)]\!]$.)

There are several aspects to the invalidity of G8a, and it's worth identifying them. For a start, we are able to perform a contraction with $\neg p$ (moving from x to y_1) *before* we can contract with $\neg(p \wedge q)$ (move from x to y_1 to y_2). As a consequence it is consistent to have $(x, y_2) \in [\![\mu\text{-con}([\sqsubseteq]\neg(p \wedge q))]\!] \setminus [\![\mu\text{-con}([\sqsubseteq]\neg p)]\!]$.[6] A related point is this. Since in our indeterministic set-up we have interpreted $\neg\psi \notin T * \varphi$ as "for *some* result of revising by φ, $\neg\psi$ is not in that result," we are able to have a revision with p that contains $\neg q$, notably z_1, while at the same time having one that does contain q. And as expansions need not always be defined in my set-up a revision with p (the move from x to y_1 to z_1) need not be a revision with $p \wedge q$.

Some of the causes underlying the invalidity of G8a can be eliminated. For example, reading $\neg\psi \notin T * \varphi$ as "for *all* results T' of revising T by φ" as in G8b, some of the indeterminism can be lifted. In particular, points like z_1 in Figure 11.2 will then be forbidden. Nevertheless, G8b is still not valid, as the reader may verify. Although one might go still furhter towards ensuring that expansions and revisions are defined when needed, I don't think that all aspects of indeterminism can be done away with. Specifically, I don't think that the kind of dependencies noted in footnotes 5 and 6 can be removed. In conclusion: there is no reasonable translation of (∗8) into DML that will make it come out valid.

I have so far tried to give a modal analysis of the Gärdenfors postulates inside DML, yielding a formal machinery for reasoning about Theory Change. The surplus value of having the full relation algebraic repertoire available in conjunction with Gärdenfors style expansion and revision operators will be discussed towards the end of this section. At this point I want to pursue the fact that one postulate, viz. (∗8), did not come out valid despite some alterations to its initial translation. This failure may prompt three reactions. One can leave things as they are, and not be bothered by the invalidity of (∗8); as (∗8) has been criticized extensively in the literature, this choice could be well argued for (cf. for example (Ryan [25])). Alternatively, one can change the rules of the game somewhat by changing the relevant postulate to one that no longer rests on the assumptions that expansions and revisions be functional and always defined. A third possibility would be to look for an alternative (modal) modelling of the postulates in DML or some other formalism. Two proposals pursuing the second option will be discussed in the following two subsections. Readers interested in alternative (modal) modellings of the Gärdenfors postulates and of postulates proposed by others are referred to (Fuhrmann [10]) and (Grahne [15]).

---

[6] As another consequence, the so-called *recovery* postulate for contraction ($T \subseteq T - \varphi + \varphi$, or in modal terms $[\sqsubseteq]\psi \rightarrow [-\varphi][+\varphi]\psi$, where $[-\varphi]$ has the obvious interpretation) is not valid in my set-up. This may not be such a bad thing as the recovery postulate is commonly considered to be the intuitively least compelling of the Gärdenfors postulates for contracting, cf. (Hansson [17]).

11.3.2   The Lindström-Rabinowitz postulates

While discussing the indeterminacy arising in the context of revision of probabilistic functions modelling belief states, one proposal Lindström and Rabinowitz [22] come up with, is letting belief revision be a relation rather than a function. They argue that this way of looking at belief revision is natural if one thinks that an agent's policies for belief change may not always yield a *unique* belief set as the result of a revision. Let a *belief revision relation* be a ternary relation $\mathcal{R}$ between knowledge sets, (consistent) formulas and knowledge sets. Lindström and Rabinowitz propose postulates $(\mathcal{R}0)$–$(\mathcal{R}4)$ below for all T, S, U and φ, ψ.

Lindström-Rabinowitz postulates for revision as a relation

$(\mathcal{R}0)$ There exists a T' such that $T\mathcal{R}_\varphi T'$.
$(\mathcal{R}1)$ If $T\mathcal{R}_\varphi S$ then $\varphi \in S$.
$(\mathcal{R}2)$ If $T \cup \{\varphi\}$ is consistent and $T\mathcal{R}_\varphi S$, then $S = T + \varphi$.
$(\mathcal{R}3)$ If $\varphi \leftrightarrow \psi$ and $T\mathcal{R}_\varphi S$, then $T\mathcal{R}_\psi S$.
$(\mathcal{R}4)$ If $T\mathcal{R}_\varphi S$, $S\mathcal{R}_\psi U$ and $S \cup \{\psi\}$ is consistent, then $T\mathcal{R}_{\varphi \wedge \psi} U$.

The intuitive reading of $T\mathcal{R}_\varphi S$ is: S is a (possible) outcome of revising T by φ. Postulate $(\mathcal{R}0)$ corresponds to the requirement that revision should be defined for all T and (consistent) φ. Postulates $(\mathcal{R}1)$–$(\mathcal{R}3)$ are the relational counterparts to the Gärdenfors postulates $(*2)$, $(*3)$ and $(*4)$, $(*6)$, and $(*8)$, respectively. Lindström and Rabinowitz don't give relational counterparts to $(*5)$ and $(*7)$. $(\mathcal{R}4)$ is new.

How can the Lindström-Rabinowitz postulates be accounted for in DML? As before we let knowledge sets be represented as sets of the form $w_\square = \{\varphi : \mathfrak{M}, w \models [\sqsubseteq]\varphi\}$. And following the definition of $[*\varphi]\psi$, the obvious choice for the relation $\mathcal{R}_\varphi$ seems to be

$$\mathcal{R}_\varphi = [\![\mu\text{-con}([\sqsubseteq]\neg\varphi); \mu\text{-exp}([\sqsubseteq]\varphi)]\!].$$

So $T\mathcal{R}_\varphi S$ iff $\exists t, s\,(T = t_\square \wedge S = s_\square \wedge (t,s) \in \mathcal{R}_\varphi)$.

Given this representation, one can reason about the revision relation $\mathcal{R}$ and its properties using the DML apparatus. For instance, idempotency properties like

$$\text{fix}(\mathcal{R}_\varphi; \mathcal{R}_\varphi)$$

can now be tested for. I leave it to the reader to check that $(\mathcal{R}0)$ fails under this representation, and that $(\mathcal{R}1)$–$(\mathcal{R}3)$ are all valid. As to $(\mathcal{R}4)$, in order to make sense of it in DML we have to decide how to represent "$S \cup \{\psi\}$ is consistent" in DML. One natural candidate is "$[\sqsubseteq]\neg\psi \notin s_\square$," where $s_\square$ represents S. But this reading does not make $(\mathcal{R}4)$ come out valid in DML. An easy counter model is given in Figure 11.3, with $T = t_\square, S = s_\square, U = u_\square, \varphi = p$ and $\psi = q$. In Figure 11.3 $(s,t) \in \mathcal{R}_p$, $(t,u) \in \mathcal{R}_q$, $s \not\models [\sqsubseteq]\neg q$, but $(s,u) \notin \mathcal{R}_{p \wedge q}$. Hence, in DML an agent has the possibility to distinguish between

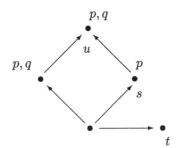

Figure 11.3
A counter model for ($\mathcal{R}$4)

revising his knowledge by φ (without excluding ψ as an unacceptable proposition) and subsequently revising by ψ on the one hand, and revising by the conjunction $\varphi \wedge \psi$ on the other hand. Thus, in DML there's still more (room for) indeterminism than is allowed for by the Lindström-Rabinowitz postulates.

11.3.3   The Katsuno-Mendelzon Postulates for Indeterministic Revision

Katsuno and Mendelzon [21] give a model-theoretic characterization of all revision operators that satisfy the Gärdenfors postulates (∗1)–(∗8). They show that these operators are precisely the ones that accomplish a revision with minimal change to the class of models of the knowledge set. This minimality is measured in terms of total pre-orders among models of the "initial" knowledge set. Katsuno and Mendelzon also study variations on the ordering notions and the corresponding postulates; in one of their variations they change the above *total* pre-orders to *partial* ones, and formulate postulates characterizing the corresponding *indeterministic* revision operators. Below I will translate these postulates into DML.

The Katsuno-Mendelzon postulates are formulated for knowledge sets T that are assumed to be represented by a propositional formula ψ_T such that $T = \{\varphi : \psi_T \vdash \varphi\}$. The notation $\psi \circ \mu$ is used to denote the revision of (the knowledge set represented by) ψ with (the formula) μ. Katsuno and Mendelzon propose seven postulates for indeterministic revision, the first five of which are in fact equivalent to the Gärdenfors postulates (∗1)–(∗7), and thus valid (when translated) in DML. Here are the remaining two.

Katsuno-Mendelzon postulates for indeterministic revision

(R7)   If $\psi \circ \mu_1$ implies μ_2 and $\psi \circ \mu_2$ implies μ_1, then $\psi \circ \mu_1$ is equivalent to $\psi \circ \mu_2$.
(R8) $(\psi \circ \mu_1) \wedge (\psi \circ \mu_2)$ implies $\psi \circ (\mu_1 \vee \mu_2)$.

Intuitively, (R7) says that if μ_2 holds in every result of revising with μ_1, and μ_1 holds in every result of revising with μ_2, then the revision with μ_1 and the revision with μ_2 have the same effect. Postulate (R8) says that every knowledge set that may be arrived at after revising with μ_1, and also after revising with μ_2, must be among the knowledge sets obtained after revising with $\mu_1 \vee \mu_2$.

Given these intuitive readings of (R7) and (R8) the following seem to be the natural translations of these postulates into DML. ($\mathcal{R}$ is the revision relation defined in the previous subsection.)

(KM7) $[*\varphi]\psi \wedge [*\psi]\varphi \rightarrow ([*\varphi]\chi \leftrightarrow [*\psi]\chi)$.

(KM8) $[*(\varphi \vee \psi)]\chi \rightarrow \neg\mathsf{do}((\mathcal{R}_\varphi \cap \mathcal{R}_\psi); \neg[\sqsubseteq]\chi?)$.

Although (*8) or G8a has now been weakened to (R7) $\wedge$ (R8) or (KM7) $\wedge$ (KM8), this weaker version is still not valid in DML. In Figure 11.4 the instance $[*p]q \wedge [*q]p \rightarrow$

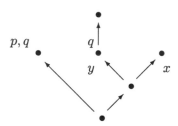

Figure 11.4
Refuting (KM7)

$([*p]r \leftrightarrow [*q]r)$ of (KM7) fails at x. As before, one thing that makes the model depicted there a counter model for (KM7) is the fact that expansions and revisions need not always be defined in my set-up. In particular, (KM7) would not fail at x in Figure 11.4 if it were possible to expand with q at y. Furthermore, in Figure 11.5 the instance $[*(p \vee q)]r \rightarrow \neg\mathsf{do}((\mathcal{R}_p \cap \mathcal{R}_q); \neg[\sqsubseteq]r?)$ of (KM8) fails at x. What this seems to amount to is that in DML an agent can get to know a (non-trivial) disjunction without having to know either disjunct. Apparently this possibility is excluded by the Katsuno-Mendelzon postulates.

A Look Back

Let's step back and review some points made in this section. One of the main features of the revision and expansion operators defined in this section as opposed to other formalisms for theory change, is that in my set-up revisions and expansions need not always be defined. Just as one can argue for giving up the functionality or determinism implied by the Gärdenfors postulates by saying that an agent's strategies for belief revision may

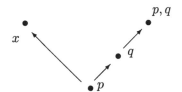

Figure 11.5
Refuting (KM8)

not always tell him how to choose between possible outcomes,—one can also argue for
the possibility of revisions and expansions not being defined at all by pointing out that
an agent's strategy for belief revision may not always tell him how to revise or expand.
Everyday life examples to this effect are easily found.

As was pointed out before, revisions and expansions as defined in this section lack
the *total* independence of sentences implicitly assumed by, for instance, the Gärdenfors
postulates for belief contraction (cf. footnotes 5, 6). This lack of independence might be
useful for modelling non-logical relations between beliefs.

Apart from the above two deviations this section shows that it is possible to define
revision and expansion operators in a fairly standard dynamic modal formalism like
DML that satisfy most of the postulates given by Gärdenfors, Lindström-Rabinowitz,
or Katsuno-Mendelzon.

There are several advantages to having revision and expansion operators satisfying
those postulates defined using the well-known Boolean and relation algebraic repertoire.
To a large extent this embedding de-mystifies the enterprise of theory change. Next, in
this larger repertoire one is no longer restricted to classical combinations of expansions
and revisions, but further operations become visible as well. One can think of sequential
composition of revisions, of reversals or 'un-doings' of revisions, and given that revisions
and expansions need not always be defined in my set-up, one might introduce *conditional*
revisions or expansions, where the conditions could read something like "after having
contracted with $\neg\varphi$ you should always be able to expand with φ."

Having the revision and expansion operations embedded in a Boolean and relation
algebraic setting also reveals possible generalizations. One might consider weaker forms
of revision in which some of the minimality requirements are weakened. Second, this
section discussed revision, that is, changing beliefs as a result of newly obtained information about a static world; one could also try and define so-called *updates* in DML;
an update is a theory change reflecting a change in the world. As shown in (Katsuno
and Mendelzon [20]) updates can be characterized by a set of postulates similar to the
Gärdenfors postulates. Another obvious generalization is to allow for several copies of
these operators, possibly interacting in certain prescribed ways, to model not only the

belief change of several agents simultaneously but also the belief changes resulting from interaction between the agents.

Below the states of DML-models will be equipped with structure, a move that could be made here as well, allowing the theories that are being revised to explicitly have structure. One can think here of a hard core of sentences not admitting revisions, surrounded by sentences which do admit revisions but which need not all have the same epistemic status; the latter kind of sentences would then be ranked according to their "epistemic entrenchment", and the revision process would need to take this into account (compare (Gärdenfors and Makinson [13])).

11.4 Terminological Languages

As Blackburn and Spaan [5] put it, in recent years modal logicians have considered a number of enriched modal systems that bear on issues of knowledge representation. One example is Schild's [26] in which the correspondence between terminological languages and modal logic is used to obtain complexity results for terminological reasoning. In this section the correspondence between DML and one particular terminological proposal will be described.

Recall that terminological languages provide a means for expressing knowledge about hierarchies of concepts. They allow the definition of concepts and roles built out of primitive concepts and roles. *Concepts* are interpreted as sets (of individuals) and *roles* are interpreted as binary relations between individuals. For instance, traveler and Amsterdam may be concepts, and has-flown-to may be a role. Compound expressions are built up using various language constructs. Quite a number of proposals for such constructs have been and still are being put forward (cf. (Schmidt [27]) for a comprehensive survey). Here I will link DML to a KL-ONE dialect discussed by Brink and Schmidt [7]; I will refer to this language as the Brink and Schmidt language.

The operations considered by Brink and Schmidt are the usual Boolean ones for the concepts plus the usual RA-operations for the roles. In addition they consider a binary operator $\Diamond$ taking a role and a concept, and returning a concept: $\Diamond(R, C) = \{ x : \exists y \, ((x, y) \in R \wedge y \in C) \}$, and a mapping $(\cdot)^c$ called *(left) cylindrification* taking concepts to roles: $C^c = \{ (x, y) : x \in C \}$. Other operations usually considered in terminological languages are *role quantifications* of the form (SOME has-flown-to amsterdam) and (ALL has-flown-to amsterdam). These expressions can be read as "objects having flown (at least once) to Amsterdam" and "objects all of whose flying trips went to Amsterdam". The quantifications (SOME R C) and (ALL R C) can be defined in Brink and Schmidt's language as $\Diamond(R, C)$ and $-\Diamond(R, -C)$, respectively.

Here's an example; while the present author is abroad one thing he may try to achieve is "writing a paper and not phoning to a Dutch person", or:

$\Diamond(\texttt{writing}, \texttt{paper}) \wedge \neg\Diamond(\texttt{phone} \cap (\texttt{dutch} \wedge \texttt{human})^c, \top)$,

where $\top$ is the Boolean 1.

The main questions in terminological reasoning are *satisfiability problems* (does a concept (role) have a non-empty denotation in some interpretation), and the *subsumption problem* (a concept (role) C subsumes a concept (role) D iff in every interpretation the denotation of C is a superset of the denotation of D). For example, on the understanding that amsterdam is in europe the concept

$$(\texttt{ALL has-traveled-to amsterdam}) \cap$$
$$(\texttt{SOME has-flown-to north-of-paris})$$

is subsumed by

$$(\texttt{ALL has-flown-to europe}).$$

As conjunction and negation are available in this languae, the subsumption problem can be reduced to the satisfiability problem.

What's the connection between Brink and Schmidt's terminological language and DML? Clearly, the terminological concepts are simply the propositions of DML, and the roles have their counterparts in the extension $DML(\{\sqsubseteq_i\}_{i \in I}, \Phi)$ of DML in which multiple 'primitive' relations $\sqsubseteq_i$ are available. So the two systems have the same basic ingredients. But what about their operators? Are they interdefinable, for example? Tables 11.2 and 11.3 show that in fact they are.

$$\Diamond(\alpha, \varphi) = \{\, x : \exists y\, ((x,y) \in [\![\alpha]\!] \wedge \mathfrak{M}, y \models \varphi)\,\} = \texttt{do}(\alpha; \varphi?),$$
$$\varphi^c = \{\, (x,y) : \mathfrak{M}, x \models \varphi\,\} = \varphi?; (\delta \cup -\delta),$$

Table 11.2
From terminological logic to DML ....

$\texttt{do}(\alpha)$	$=$	$\Diamond(\alpha, 1),$	$\texttt{exp}(\varphi)$	$=$	$\sqsubseteq \cap \varphi^c$,
$\texttt{ra}(\alpha)$	$=$	$\Diamond(\alpha, 1),$	$\texttt{con}(\varphi)$	$=$	$\sqsubseteq \cap (\neg\varphi)^c$,
$\texttt{fix}(\alpha)$	$=$	$\Diamond((\alpha \cap \delta), 1),$	$\varphi?$	$=$	$\delta \cap \varphi^c.$

Table 11.3
... and conversely.

To illustrate this connection, here's an example expressing the concept "people having flown only to cities called Amsterdam" in DML:

$$\texttt{human} \wedge \texttt{do}(\texttt{has-flown-to}; (\texttt{city} \wedge \texttt{amsterdam})?) \wedge$$
$$\neg\texttt{do}(\texttt{has-flown-to}; -(\texttt{city} \wedge \texttt{amsterdam}?)).$$

The above connection may be formulated 'officially' by means of two mappings between the two languages, thus establishing the following result.

PROPOSITION 11.1 Brink and Schmidt's language for terminological reasoning with primitive concepts Φ and primitive roles $\{\sqsubseteq_i\}_{i\in I}$ is a notational variant of the modal language $DML(\{\sqsubseteq_i\}_{i\in I}, \Phi)$.

Thus, the main issues in terminological reasoning, viz. satisfiability and subsumption, may be re-formulated as satisfiability problems in (an extension of) DML, and results and topics from the modal domain can be transferred to the terminological domain, and vice versa. To substantiate this claim, let me give some examples.

COROLLARY 11.1 Modulo the translation induced by Table 11.3, the axioms and inference rules of $DML(\{\sqsubseteq_i\}_{i\in I}, \Phi)$ are a sound and complete axiomatization of subsumption of concepts in the Brink and Schmidt language.

We can be very brief about the proof of Corollary 11.1: apply 11.1 and 11.1. Indirectly, the axioms and rules also of $DML(\{\sqsubseteq_i\}_{i\in I}, \Phi)$ also axiomatize subsumption of *roles* in the Brink and Schmidt language; this is because any equation $\alpha = \beta$ between roles (relations) can be mimicked at the level of concepts (propositions) by

$$EQUAL(\alpha, \beta) := \mathbf{A}(\neg\mathsf{do}(\alpha \cap -\beta) \wedge \neg\mathsf{do}(-\alpha \cap \beta)).$$

Although the following result is not new (cf. Schmidt-SchauS [28]), its proof too comes very easy given Proposition 11.1, and the fact that satisfiability in DML is undecidable (by De Rijke [23, Theorem 5.1]).

COROLLARY 11.2 Satisfiability and subsumption in the Brink and Schmidt language are undecidable.

As is well known, part of the Knowledge Representation community is concerned with finding *tractable* terminological systems, either by limiting the expressive power of the representation language, or by limiting the inference capabilities of the formalisms. This has resulted in the description of quite a number of decidable or even tractable systems, many of which can be seen as fragments of the Brink and Schmidt system. By 11.1, this work is relevant to the search for decidable or tractable fragments of DML.

Here's a final possibility for exchange between the modal and terminological domain. Terminological reasoning often deals with *number restrictions* like (≥ 2 has-flown-to amsterdam) (which can be read as "objects having flown to Amsterdam at least twice") to perform numerical comparisons. The modal logic of these counting expressions (by themselves) has been analyzed by Van der Hoek and De Rijke [19]. The link between terminological languages and DML established in 11.1 suggests that it may be worth the effort to add the counting quantifiers to DML, and examine the resulting language.

To finish this section let me cast the connection between the Brink and Schmidt language and DML in algebraic terms. Schmidt [27] equips the Brink and Schmidt language with an algebraic semantics called Peirce algebras. To understand what these are we have to go through one or two definitions. First of all, a *Boolean module* is a structure $\mathfrak{M} = (\mathfrak{B}, \mathfrak{R}, \Diamond)$, where $\mathfrak{B}$ is a Boolean algebra, $\mathfrak{R}$ is a relation algebra and $\Diamond$ is a mapping $\mathfrak{R} \times \mathfrak{B} \to \mathfrak{B}$ such that

M1 $\Diamond(r, a + b) = \Diamond(r, a) + \Diamond(r, b)$ M4 $\Diamond(\delta, a) = a,$

M2 $\Diamond(r + s, a) = \Diamond(r, a) + \Diamond(s, a)$ M5 $\Diamond(0, a) = 0,$

M3 $\Diamond(r, \Diamond(s, a)) = \Diamond((r; s), a)$ M6 $\Diamond(r, \Diamond(r, a)') \leq a'.$

Just as Boolean algebras formalize reasoning about sets, and relation algebras formalize reasoning about relations, Boolean modules formalize reasoning about sets interacting with relations through $\Diamond$. In the full Boolean module $\mathfrak{M}(U) = (\mathfrak{B}(U), \mathfrak{R}(U), \Diamond)$ over a set $U \neq \emptyset$ the operation $\Diamond$ is defined as described earlier, by

$$\Diamond(R, C) = \{\, x : \exists y\, ((x, y) \in R \wedge y \in C) \,\}.$$

(See (Brink [6]) for a formal definition of Boolean modules and some examples.) Boolean modules are almost, but not quite, the right algebraic semantics for Brink and Schmidt's terminological language. To obtain a perfect match, what we need in addition to the *set* forming operation $\Diamond$, is an operation that forms new *relations* out of sets. This yields the notion of a *Peirce algebra*, which is a two-sorted algebra $\mathfrak{P} = (\mathfrak{B}, \mathfrak{R}, \Diamond, (\cdot)^c)$ with $(\mathfrak{B}, \mathfrak{R}, \Diamond)$ a Boolean module, and $(\cdot)^c : \mathfrak{B} \to \mathfrak{R}$ a mapping such that for every $a \in \mathfrak{B}$, $r \in \mathfrak{R}$ we have

P1 $\Diamond(a^c, 1) = a,$

P2 $\Diamond(r, 1)^c = r; 1.$

In the full Peirce algebra $\mathfrak{P}(U)$ over a set $U \neq \emptyset$, $(\cdot)^c$ is defined, as before, as $C^c = \{\, (x, y) : x \in C \,\}$. The algebraic apparatus of Peirce algebras has been used by Brink and Schmidt [7] as an inference mechanism in terminological representation

Where does DML come in here? Because of Proposition 11.1, the modal algebras for the DML-language $L(\{\sqsubseteq_i\}_{i \in I}, \Phi)$ are the Peirce algebras generated by the relations $\{\sqsubseteq_i\}_{i \in I}$ and the propositions Φ. Let a *set identity* be one of the form $a = b$ where a, b are terms living in the Boolean reduct of a Peirce algebra (observe that a, b may contain the relational operations as well as $\Diamond$ and $(\cdot)^c$). Then, the completeness result 11.1 may be interpreted as follows.

PROPOSITION 11.2   The set identities valid in all representable Peirce algebras are completely axiomatized by the algebraic counterpart of the modal axiom system for the language $DML(\{\sqsubseteq_i\}_{i \in I}, \Phi)$.[7]

---

[7]Although, strictly speaking, the completeness result 11.1 only axiomatizes validity on pre-ordered DML-structures, the construction does not depend in an essential way on these structural assumptions.

Axiomatic aspects of the full two-sorted language over Peirce algebras, invlobing both set identities and identities between terms denoting relations, are studied by De Rijke [24].

11.5   Structured States: Update Semantics

In both this and the next section I will equip the states of DML-models with additional structure to be able to link DML with other dynamic proposals. The formalism I will consider in the present section is Veltman's *update semantics* [30]. In this system the standard explanation of the meaning of a sentence being its truth-conditions, is replaced by: "you know the meaning of a sentence if you know the changes it brings about in the information state of anyone who accepts the information conveyed by the sentence." According to this point of view the meaning of a sentence becomes a dynamic notion, an operation on information states. In this dynamic approach phenomena surrounding the instability and changing of information caused by modal qualifications like 'might,' 'presumably' and 'normally' can be adequately accounted for, as is shown by Veltman using a number of systems. The simplest one, called US_1 here, has in its vocabulary a unary operator might and a connective '•', in addition to the usual Boolean connectives; in US_1 one can reason about an agent acquiring new information about the actual facts.

DEFINITION 11.1   The language of $US_1(\Phi)$ is given by the following definition.

$$
\begin{array}{ll}
\text{Atomic formulas:} & p \in \Phi, \\
\text{Simple formulas:} & \varphi \in Form_1(\Phi), \\
\text{Formulas:} & \psi \in US_1(\Phi).
\end{array}
$$

$$
\begin{aligned}
\varphi &::= p \mid \bot \mid \neg\varphi \mid \varphi_1 \vee \varphi_2 \mid \texttt{might}\,\varphi, \\
\psi &::= \varphi \mid \psi_1 \bullet \psi_2.
\end{aligned}
$$

The important restriction is that no • is allowed to occur in the scope of a might.

The intuitive reading of might φ is that one has to agree to might φ if φ is consistent with one's knowledge; otherwise might φ is to be rejected. The operator • is simply the composition of (the functions expressed by) formulas.

DEFINITION 11.2   The semantics of the update system US_1 is as follows. Let $W \subseteq 2^\Phi$; a subset of W is an *information state*. Formulas are interpreted as functions in $2^W \longrightarrow 2^W$, that is, as functions from information states to information states. Let $\sigma \subseteq W$. I write $[\varphi]\sigma$ for the result of updating σ with φ.

$$\begin{aligned}
[p]\sigma &= \sigma \cap \{w : p \in w\}, \\
[\neg\varphi]\sigma &= \sigma \setminus [\varphi]\sigma, \\
[\varphi \vee \psi]\sigma &= [\varphi]\sigma \cup [\psi]\sigma, \\
[\text{might } \varphi]\sigma &= \begin{cases} \sigma, & \text{if } [\varphi]\sigma \neq \emptyset \\ \emptyset, & \text{otherwise,} \end{cases} \\
[\varphi \bullet \psi]\sigma &= [\psi]([\varphi]\sigma).
\end{aligned}$$

Veltman discusses several notions of valid consequence. Since these are not my prime concern here (but see below), I will confine myself to explaining the notion "$US_1 \models \varphi$" as "for all information states σ, $[\neg\varphi]\sigma = \emptyset$."

Van Eijck and De Vries [8] have established a connection between US_1 and the modal system $S5$ (see also [16]). This construction underlies the embedding of US_1 into DML presented below; what it amounts to is that US_1 is a formalism for reasoning about $S5$-models and certain transitions between them. This inspires the following definition.

DEFINITION 11.3   A *structured DML-model* is a tuple $\mathfrak{M} = (W_g, \sqsubseteq, \llbracket \cdot \rrbracket)$ where $\sqsubseteq$ is a global relation on W_g, and W_g is a set of (finite) *pointed $S5$-models* of the form $\mathfrak{m} = (W, R, w, V)$ such that $w \in W$, $R = W \times W$, and V is a valuation. Moreover, the following conditions should be satisfied:

- $\mathfrak{m} \sqsubseteq \mathfrak{n}$ iff $\mathfrak{m}$ is a submodel of $\mathfrak{n}$,
- if $\mathfrak{m} = (W, R, w, V) \in \mathfrak{M}$ then $(W, R, v, V) \in \mathfrak{M}$ for all $v \in W$, and $\mathfrak{n} \in \mathfrak{M}$ for all $\mathfrak{n} \sqsubseteq \mathfrak{m}$.

The formal language appropriate for reasoning about such 2-level structures, $DML(S5)$, is defined as follows. Starting from a set of proposition letters Φ, $S5$-formulas are built using the operators L, M in the usual way. Let Φ' be the resulting set; this is the set of *local* formulas. They serve as 'proposition letters' for the global language DML; that is: $DML(S5)$ formulas are obtained from Φ' by applying the usual DML-connectives to its elements.

The important semantic clauses then read as follows, for $\mathfrak{m} = (W, R, w, V)$:

$\mathfrak{M}, \mathfrak{m} \models p$ iff $w \in V(p)$

$\mathfrak{M}, \mathfrak{m} \models M\varphi$ iff   for some $v \in W$ with wRv, $(W, R, v, V) \models \varphi$,

that is, the value of such formulas is computed locally. For 'purely global' formulas, on the other hand, the value is computed globally, as in the following example:

$\mathfrak{M}, \mathfrak{m} \models \text{do}(\sqsubseteq)$ iff   for some $\mathfrak{n} \in \mathfrak{M}$, $\mathfrak{m} \sqsubseteq \mathfrak{n}$.

DEFINITION 11.4 Define a translation $(\cdot)^\dagger$ of the US_1-language into $DML(S5)$ as follows:

$$
\begin{aligned}
(p)^\dagger &= p \\
(\neg\varphi)^\dagger &= \neg\varphi^\dagger \\
(\varphi \vee \psi)^\dagger &= \varphi^\dagger \vee \psi^\dagger \\
(\mathtt{might}\,\varphi)^\dagger &= M\varphi^\dagger \\
(\varphi \bullet \psi)^\dagger &= \varphi^\dagger \wedge \mathsf{do}\Big([\sqsupseteq;(L\varphi^\dagger \wedge \psi^\dagger)?] \\
&\quad \cap - [(\sqsupseteq;(L\varphi^\dagger \wedge \psi^\dagger)?);((\sqsupseteq \cap - \delta);(L\varphi^\dagger \wedge \psi^\dagger)?)]\Big).
\end{aligned}
$$

The intuitive reading of $(\mathtt{might}\,\varphi)^\dagger$ is that we *locally* check whether there is a point verifying $\varphi^\dagger$. The intuitive interpretation of $(\varphi \bullet \psi)^\dagger$ is that $(\varphi \bullet \psi)^\dagger$ holds at $\mathsf{m} = (W, R, w, V)$ iff $\mathsf{m} \models \varphi^\dagger$ and for $S = \{x \in \mathsf{m} : (W, R, x, V) \models \varphi^\dagger\}$ we have that $(S, S^2, w, V \restriction S) \models \psi^\dagger$. Notice that $(\cdot)^\dagger$ takes US_1-formulas into a *decidable* fragment of $DML(S5)$.

PROPOSITION 11.3 Let φ be a formula in $US_1(\Phi)$. Then we have $US_1 \models \varphi$ iff $DML(S5) \models \varphi^\dagger$.

Proof Suppose $US_1 \not\models \varphi$. Then for some $W \subseteq 2^\Phi$, and $\sigma \subseteq W$, $[\neg\varphi]\sigma \neq \emptyset$. Define $\mathfrak{M} = (W_g, \sqsubseteq, [\![\cdot]\!])$, where $W_g =$

$$\{ (\sigma, \sigma^2, x, V_\sigma) : \sigma \subseteq W, x \in \sigma, (p \in V_\sigma(y) \text{ iff } p \in y, \text{ for } y \in \sigma) \},$$

and $\sqsubseteq$ and $[\![\cdot]\!]$ have their standard interpretation. Then, by a simple formula induction, we have that $\forall\psi\forall\sigma \subseteq W\forall j \in \sigma\,(j \in [\psi]\sigma$ iff $(\sigma, \sigma^2, j, V_\sigma) \models \psi)$. It follows that $DML(S5) \not\models \varphi^\dagger$.

To prove the opposite direction we proceed as follows. Assume that for some $\mathfrak{M}$ and $\mathsf{m} \in \mathfrak{M}$ we have $\mathfrak{M}, \mathsf{m} \not\models \varphi^\dagger$. Let $\mathsf{m} = (W, W^2, w, V_\mathsf{m})$. By standard modal logic we may assume that in m every 'relevant' state description of the form $(\neg)p_0 \wedge (\neg)p_1 \wedge \ldots \wedge (\neg)p_n$ (where $p_0, \ldots, p_n$ are all the proposition letters occurring in φ), occurs only once in m. We may also assume that for every $\mathsf{n} = (W_1, R_1, w_1, V_1) \in \mathfrak{M}$, (W_1, R_1, V_1) is a substructure of (W, W^2, V). Now, let $W' = 2^{\{p_0, \ldots, p_n\}}$, and for $\mathsf{n} \in \mathfrak{M}$ let $\sigma_\mathsf{n} \subseteq W$ be the set of state descriptions realized in n. Then, by a simple inductive proof, we have for all formulas ψ containing at most the proposition letters $p_0, \ldots, p_n$, and all $\mathsf{n} = (W_1, R_1, w_1, V_1) \in \mathfrak{M}$, $\mathsf{n} \models \psi^\dagger$ iff $w_1 \in [\psi]\sigma_\mathsf{n}$, which completes the proof.

Proposition 11.3 may be interpreted as saying that the 'internal' notions of US_1 can be turned into internal notions of DML. But some of the 'external' or meta-notions of

US_1 can also be turned into internal notions of DML. Veltman [30] discusses various notions of valid consequence for his update systems, including

$$\varphi_1, \ldots, \varphi_n \models_1 \psi \quad \text{iff} \quad \text{for all } \sigma \text{ such that } [\varphi_i]\sigma = \sigma,$$
$$\text{we have } [\psi]\sigma = \sigma \ (1 \le i \le n), \text{ and}$$
$$\varphi_1, \ldots, \varphi_n \models_2 \psi \quad \text{iff} \quad \text{iff for all } \sigma, \text{ we have that}$$
$$[\psi]([\varphi_n](\ldots([\varphi_1]\sigma)\ldots)) = [\varphi_n](\ldots([\varphi_1]\sigma)\ldots).$$

In $DML(S5)$ these notions become

$$\varphi_1, \ldots, \varphi_n \models_1 \psi \quad \text{iff} \quad \mathbf{A}(L\varphi^\dagger \wedge \ldots \wedge L\varphi_n^\dagger \to L\psi^\dagger), \quad \text{and}$$
$$\varphi_1, \ldots, \varphi_n \models_2 \psi \quad \text{iff} \quad \mathbf{A}L\big((\varphi_1 \bullet \cdots \bullet \varphi_n)^\dagger \to \psi^\dagger\big),$$

respectively, where $\mathfrak{M}, \mathfrak{m} \models \mathbf{A}\chi$ iff for all $\mathfrak{n} \in \mathfrak{M}$ we have $\mathfrak{M}, \mathfrak{n} \models \chi$.

Given the embedding of US_1 into DML, some natural extensions and generalizations become visible. Besides **might** one can consider other tests, the most obvious of which are definable in DML, like an operator testing whether updating with a formula φ will change the current state, or one testing whether the current state is at all reachable via an update with φ, or whether a pre-given goal state may be reached by performing certain updates.

11.6 Structured States: Preferences

Among the structures of logic L there may be some models that are *preferred* for one reason or another. Preferences may differ between applications, thus giving rise to different notions of preferential inference. Shoham [29] offers a general to preferential reasoning in which there is a (strict partial) order of preference $<$ on L-models on top which minimal consequence is defined as "truth of the conclusion in all $<$-minimal or most-preferred L-models of the premises." By specifying the relation $<$ in alternative ways many formalisms with non-monotonic aspects can be shown to fit this general preferential scheme.

Given the embedding of US_1 into DML of 11.5 as an example, it should be obvious how preferential reasoning can be mimicked in DML: let $\sqsubseteq$ be the preferential ordering, and let the states of our DML-models simply *be* L-models. Then 'φ preferentially entails ψ' is true in the global structure $\mathfrak{M}$ iff $\mathfrak{M} \models \mathbf{A}(\varphi \wedge \neg \mathsf{do}(\sqsubseteq \ ; \varphi?) \to \psi)$. Via this equivalence all preferential reasoning can be performed inside DML.

Just as the preference relation embodies certain dynamic aspects of the underlying L, it itself could also be subjected to change. This point may be illustrated with a system US_2 which is slightly more complex than US_1, and which has also been introduced by

Veltman [30]. In US_2 one is not only able to reason about changing knowledge as new information comes in, but also about changing expectations; the latter are modelled using a notion of *optimality* with respect to a pre-order. Modelling this system in DML requires adding a separate $S4$-like component for expectations to the structured states of 11.5, in addition to the $S5$-like component for knowledge. An agents refinement or revision of his expectations can then be modelled inside such DML-structures by making moves to points with a suitably altered 'expectations' component.

11.7 Final Remarks

Let me point out what I consider to be the main points of this chapter. It has brought out connections and analogies between dynamic formalisms from cognitive science, philosophy and computer science by using a fairly traditional dynamic modal system (DML) in a flexible way, far beyond its traditional boundaries.

Putting DML to work in this manner had the surplus advantage of de-mystifying some of those formalisms, and through these applications natural alternatives and generalizations of formalisms in those areas became visible.

Finally, structuring states as in 11.5 and 11.6 of this note may be seen as initial steps of a larger program of adding structure to objects. As to adding structuring the transitions between states, rather than or in addition to structuring the states, there seems to be a problem. When transitions are equated with pairs of objects rather than treated as first-class citizens in their own right, there does not seem to be an obvious way to structure them. But Van Benthem [3] proposes a system of *arrow logic* in which the transitions or arrows have a primary status in the ontology, without necessarily being identified with pairs of states. Eventually this might be the way to go if one wants to be able to structure transitions as well as objects.

Acknowledgments

This chapter was conceived of during a visit to the Center for the Study of Language and Information, Stanford University, on NWO-grant SIR 11-596. I would like to thank David Israel for valuable suggestions. Back in the Netherlands Wiebe van der Hoek, Frank Veltman and especially Johan van Benthem provided useful comments.
The investigations were supported by the Foundation for Philosophical Research (SWON), which is subsidized by the Netherlands Organization for Scientific Research (NWO).

References

[1]   Alchourrón, C., Gärdenfors, P. & Makinson, D. (1985), 'On the logic of theory change: partial meet contraction and revision functions', *Journal of Symbolic Logic* **50**, 510–530.

[2]   van Benthem, J. (1989), Semantic parallels in natural language and computation, *in* H.-D. Ebbinghaus et al., eds, 'Logic Colloquium. Granada 1987.', North-Holland, Amsterdam, pp. 331–375.

[3]   van Benthem, J. (1991), *Language in Action*, North-Holland, Amsterdam.

[4]   van Benthem, J. (1991), Logic and the flow of information, *in* D. Prawitz et al., eds, 'Proc. 9th Intern. Congress of Logic, Method. and Philos. of Science', North-Holland, Amsterdam.

[5]   Blackburn, P. & Spaan, E. (1993), 'A modal perspective on the computational complexity of attribute value grammar', *Journal of Logic, Language and Information*. To appear.

[6]   Brink, C. (1981), 'Boolean modules', *Journal of Algebra* **71**, 291–313.

[7]   Brink, C. & Schmidt, R. (1992), 'Subsumption computed algebraically', *Computers Math. Applic.* **23**, 329–342.

[8]   van Eijck, J. & de Vries, F.-J. (1993), Reasoning about update logic, Technical Report CS-R9312, CWI, Amsterdam. To appear in the *Journal of Philosophical Logic*.

[9]   Finger, M. & Gabbay, D. (1992), 'Adding a temporal dimension to a logic system', *Journal of Logic, Language and Information* **1**, 203–233.

[10]   Fuhrmann, A. (1990), On the modal logic of theory change, *in* A. Fuhrmann & M. Morreau, eds, 'Lecture Notes in AI, 465', Springer, pp. 259–281.

[11]   Gabbay, D., Hodkinson, I. & Reynolds, M. (1993), *Temporal Logic: Mathematical Foundations and Computational Aspects*, Oxford University Press. To appear.

[12]   Gärdenfors, P. (1988), *Knowledge in Flux*, The MIT Press, Cambridge, MA.

[13]   Gärdenfors, P. & Makinson, D. (1988), Revisions of knowledge systems using epistemic entrenchment, *in* 'Proc. 2nd. Conf. Theoret. Aspects of Reasoning about Knowledge', Pacific Grove, pp. 83–95.

[14]   Gargov, G. & Passy, S. (1990), A note on Boolean modal logic, *in* P. Petkov, ed., 'Mathematical Logic. Proceedings of the 1988 Heyting Summerschool', Plenum Press, New York, pp. 311–321.

[15]   Grahne, G. (1991), Updates and counterfactuals, *in* J. Allen, R. Fikes & E. Sandewall, eds, 'Principles of Knowledge Representation and Reasoning: Proc. 2nd. Intern. Conf.', Morgan Kaufman, San Mateo, pp. 269–276.

[16]   Groeneveld, W. (1992), Contextual $S5$: a 'static' renotation of Update Semantics (Draft), Manuscript.

[17]   Hansson, S. (1991), 'Belief contraction without recovery', *Studia Logica* **50**, 251–260.

[18]   Harel, D. (1984), Dynamic logic, *in* D. Gabbay & F. Guenthner, eds, 'Handbook of Philosophical Logic', Vol. 2, Reidel, Dordrecht, pp. 497–604.

[19]   van der Hoek, W. & de Rijke, M. (1993), 'Generalized quantifiers and modal logic', *Journal of Logic, Language and Information* **2**, 19–58.

[20]   Katsuno, H. & Mendelzon, A. (1991), On the difference between updating a knowledge base and revising it, *in* J. Allen, R. Fikes & E. Sandewall, eds, 'Principles of Knowledge Representation and Reasoning: Proc. 2nd Intern. Conf.', Morgan Kaufman, San Mateo, pp. 387–394.

[21]   Katsuno, H. & Mendelzon, A. (1992), 'Propositional knowledge base revision and minimal change', *Artificial Intelligence* **52**, 263–294.

[22]   Lindström, S. & Rabinowitz, W. (1989), 'On probabilistic representation of non-probabilistic belief

revision', *Journal of Philosophical Logic* **18**, 69–101.

[23]   de Rijke, M. (1992), A system of dynamic modal logic, Technical report # CSLI-92-170, Stanford University. Also appeared as ILLC Report LP-92-08, University of Amsterdam.

[24]   de Rijke, M. (1993), Modal axioms for Peirce algebras, Manuscript, ILLC, University of Amsterdam.

[25]   Ryan, M. (1992), Ordered Theory Presentations, PhD thesis, Imperial College.

[26]   Schild, K. (1990), A correspondence theory for terminological logics, *in* 'Proc. 12th IJCAI', Sydney, Australia, pp. 466–471.

[27]   Schmidt, R. (1991), Algebraic terminological representation, Master's thesis, Department of Mathematics, University of Cape Town.

[28]   Schmidt-SchauS, M. (1989), Subsumption in KL-ONE is undecidable, *in* 'Proc. 1st Intern. Conf. Princ. of Knowledge Representation and Reasoning', Morgan Kaufman, San Mateo, pp. 421–431.

[29]   Shoham, Y. (1987), Nonmonotonic logics: meaning and utility, *in* 'Proc. IJCAI-87', Milan, Italy, pp. 388–393.

[30]   Veltman, F. (1992), 'Defaults in update semantics', *Journal of Philosophical Logic*. To appear.

12 Actions under Presuppositions

Albert Visser

12.1 Introduction

This chapter consists of two parts: the first is contained in section 12.2. It reviews, perhaps too briefly, some basic philosophy on meaning, information, information state, information ordering and the like. In the remaining sections two interwoven problems are considered: the first is how to view update functions as partial states (or more generally partial 'actions'). Partial states (actions) are viewed as states (actions) under a presupposition. The second problem is how the merger of meanings interacts with the synchronic information ordering. We explore some consequences of the hypothesis that this interaction is described by a Residuation Lattice.

Es kommt mir darauf an zu zeigen, daS das Argument nicht mit zur Funktion gehört, sondern mit der Funktion zusammen ein vollständiges Ganzes bildet; denn die Funktion für sich allein is unvollständig, ergänzungsbedürftig oder ungesättigt zu nennen. Und dadurch unterscheiden sich die Funktionen von den Zahlen von Grund aus. (Frege [8, 21–22]).

Partial actions and states, that is what this chapter is about. A partial action is viewed as something unsaturated or etwas ergänzungsbedürftiges. We model the partial objects as certain partial update functions which can in their turn be represented by pairs of total objects. The first component of such a pair can be seen as a test: if you satisfy the test you can plug the hole. Consequently you can benefit from the resulting total object. This total object is given by the second component. As we shall see, this construction is quite similar to the familiar construction of the integers from the natural numbers.

12.2 Pictures & Updates

This section is an attempt to place the chapter in the somewhat broader, but hazy context of research in dynamic and discourse semantics. Some notations and basic notions are introduced.

12.2.1 Setting the Stage

Consider two simple-minded pictures of *meaning* or, perhaps, *information content*. The first is the DRT-view (DRT = Discourse Representation Theory; it is primarily due to Heim and Kamp): a meaning is like a picture, is like a structured database, is like a mental state. These meanings are called DRS's (= Discourse Representation Structures). The second is the imagery of Update Semantics (primarily due to Gardenförs and Landman &

Veltman): a meaning or information item is (or can be represented as) an update function of mental states. The purpose of this chapter is to study the relationship between these pictures (or more accurately: certain aspects of this relationship).

Prima facie these views are quite different. The DRT-view provides static objects, while the essence of the Update picture is meaning-as-something-dynamic. Also there must be far more updates of mental states, than there are mental states. Our basic idea to resolve the tension between the two views is (i) to consider only a restricted class of update functions and (ii) to represent these functions as *partial states* or *states under a presupposition*. The original states or total states are embedded in a natural way among their partial brethren.

12.2.2 Monoids for Merging Meanings

Databases or pictures can be put together or merged. Update functions can be composed. In both cases there is a fundamental operation: the *merger* respectively *function composition*. These operations are associative. We stipulate the presence of an unit element 1 for these operations. The identity is the empty database respectively the identity function. We assume that 1 is a (total) state: the state of absolute ignorance or *tabula rasa*.

We use the expression *merger* for whatever basic associative function glues meanings together, thus viewing function composition as a special case of the merger.

We take the merger as in some sense the basic or fundamental operation on meanings. Other operations are either defined in terms of it or in some wider sense derived from it. Of course, in the light of the generality of our present discussion, taking the merger as fundamental is only a *schematic* step. Yet it serves already to distinguish the present approach from Montagovian Semantics, where the basic operation is Function Application.

In view of the foregoing discussion we see that meanings form a *monoid* $\mathcal{M} = \langle M, \cdot, 1 \rangle$. We take the mental states to be a subset of the meanings. Thus we define: a *merge algebra* $\mathcal{M}$ is a structure $\langle M, S, \cdot, 1 \rangle$, where $1 \in S \subseteq M$ and where $\langle M, \cdot, 1 \rangle$ is a monoid.

CONVENTIONS 12.2.2.1

We let x, y, z, ... range over M and s, t, ... over S.

We use postfix notation for function application. Our notation for function composition is in line with this convention: $xF \circ G := (xF)G$.

If F and G are partial functions we write e.g. $wF = uG$ for: either sF and uG are both defined and have the same value, or both are undefined. We write $sF \cong uG$ for sF and uG are both defined and their values are equal.

We associate update functions to our algebra in the obvious way. An update function on a merge algebra $\mathcal{M}$ is a partial function from S to S. To each x in M we associate an update function Φ_x as follows:

$s\Phi_x := s \cdot x$ if $s \cdot x \in S$, $s\Phi_x$ is undefined otherwise.

We say that $\mathcal{M}$ is an *update algebra* if the map Φ with $x\Phi = \Phi_x$ is a homorphism from $\langle M, \cdot, 1 \rangle$ to $\langle$the update functions on $M, \circ, ID \rangle$, where ID is the identity function on S.

Under what conditions is a merge algebra an update algebra? The answer is that the algebra has to satisfy the $OTAT$-principle.

DEFINITION 12.1 (THE $OTAT$-PRINCIPLE) $\mathcal{M}$ satisfies $OTAT$ if:

for all $x, y \in M$: $x \cdot y \in S \Longrightarrow x \in S$.

$OTAT$ means: once a thief, always a thief. If something fails to be a state, then it will never become a state whatever happens afterwards.

There is also a local version of $OTAT$. An element y of M has the $OTAT$ property iff for all $x \in M$: $x \cdot y \in S \Longrightarrow x \in S$.

We have the following theorem.

THEOREM 12.1 $\mathcal{M}$ is an update algebra iff $\mathcal{M}$ satisfies $OTAT$.

The theorem is an immediate corollary of:

LEMMA 12.1   Consider $y \in M$

$\forall s \in S \, \forall x \in M \; s\Phi_x\Phi_y = s\Phi_{x \cdot y} \leftrightarrow y$ has the $OTAT$ property.

Proof "$\Longrightarrow$" Suppose $\forall s \in S \, \forall x \in M \; s\Phi_x\Phi_y = s\Phi_{x \cdot y}$. Consider any x in M and suppose $x \cdot y \in S$. It follows that $1\Phi_{x \cdot y}$ is defined and hence (ex hypothesi) so is $1\Phi_x\Phi_y$. Thus $1\Phi_x$ must be defined, which means that $x \in S$.

"$\Longleftarrow$" Suppose y has the $OTAT$-property. Consider any s and y. If $s\Phi_x\Phi_y$ is defined then $s\Phi_x\Phi_y = s \cdot x \cdot y$ and $s \cdot x \cdot y \in S$. Hence $s\Phi_{x \cdot y}$ is defined and $s\Phi_{x \cdot y} = s \cdot x \cdot y$. Conversely suppose $s\Phi_{x \cdot y}$ is defined. In this case $s\Phi_{x \cdot y} = s \cdot x \cdot y$ and $s \cdot x \cdot y \in S$. By the $OTAT$ property we find that $s \cdot x \in S$ and hence $s\Phi_x$ is defined. Since $s\Phi_x = s \cdot x$ also $s\Phi_x\Phi_y$ is defined and $s\Phi_x\Phi_y = s \cdot x \cdot y$. ∎

The $OTAT$-principle suggests that the partiality of the non-states is something backwards looking, a kind of lack on the input side rather than at the output side. In other words the $OTAT$-principle suggests that meanings have a *presuppositional structure*.

12.2.3   The True and the Proper Nature of States

What is a state? The connotations of the word *state* suggest that a state is something static. Thus the notion of state would have its proper place in the static-dynamic opposition. A state would be something like a *test* or a *condition*.

I disagree with this idea. First etymology can be misleading. Think of e.g. *state of motion* versus *state of rest* to illustrate that even if a state is a something-a-thing-is-in-at-a-particular-moment, a state is not necessarily something that has no 'active' properties. A state may contain the germs of the next state and may even be said to be one of the causes of the next state. Secondly the *OTAT*-principle suggests that states find their natural home within the saturated-unsaturated distinction. A state is something that is saturated towards the past, i.o.w. something that carries no presupposition.

Let's consider some examples. In our batch of examples we choose to ignore all possible sources of unsaturatedness, other than those arising from anaphoric phenomena (at the surface level).

i.   A dog comes in.

ii. It barks.

iii. All dogs bark.

iv. A dog comes in. It barks at me.

I would say that (the meaning of/the content of) (i) is a state. No referents need to be supplied from previous discourse.

Of course calling (i) a state, carries the suggestion that (i) could be the whole knowledge-content of an organism. I'm inclined to think that if *logical possibility* is intended here, this is true. On the other hand such radical claims are not really at issue in the present discussion. We could easily stipulate that *state* here is intended as part or aspect of the total holistic (*pardonnez le mot*) state of an organism, that can be considered as standing on its own relative to a certain kind of analysis carrying its own degree of resolution (etc.).

Note that (i) is saturated when seen from the past, but not so when seen from the future—traveling for a short moment backwards in time—, since it exports a referent to later discourse. (i) is not a test or condition, since it exports a *new* referent to later discourse.

(ii) is not a state since it is unsaturated or, with Frege's beautiful expression, *Ergänzungsbedürftig* towards the past (and in this case the future too). On the other hand (ii) is static, since it neither creates nor destroys a referent. Thus we may say that (ii) is a test (for barkiness).

We leave it to the industrious reader to see that (iii) is both a state and a test and that (iv) is neither state nor test.

EXCURSION 12.2.3.1 Let's briefly glance at some possible paraphrases of (i)–(iv) in the language of DRT/DPL (DPL = Dynamic Predicate Logic, a variant of Predicate Logic introduced by Groenendijk & Stokhof):

pi. $\exists x.DOG(x).COME\text{-}IN(x)$

pii. $BARK(x)$

piii. $\forall x\,(DOG(x) \rightarrow BARK(x))$ $\big($or: $(\exists x.DOG(x) \rightarrow BARK(x))\big)$

piv. $\exists x.DOG(x).COME\text{-}IN(x).SPEAKER(y).BARK\text{-}AT(x,y)$

In (pi) the quantifier $\exists x$ introduces a new variable, but unlike in ordinary Predicate Logic the (possible) scope of this variable is not constrained to the formula given. We can go on and 'merge' (the "." stands for the merger here) e.g. $BARK(x)$ with the formula (pi). The variable x will in this case get bound. The whole formula (pi) functions as a quantifier where values of the variable x are constrained to incoming dogs. Thus x is *bound*, but *active*. A symptom of this phenomenon is that α-conversion does not preserve meaning here. On the other hand x in (pii) is not bound, but still active: x 'asks' for a value to be imported from previous discourse and sends this value on to later discourse. The variable x in the formula (piii) is bound but not active. Thus it is a classical bound variable as in Predicate Logic. The reader is invited to draw her own conclusions on (piv).

A variable occurrence is bound in a formula if it is not 'visible' from the past, like x in $\exists x.A(x)$. Dually a variable is *trapped* if it is not visible from the future, travelling backwards in time. Examples here would be the occurrence of x in $\forall x\,(A(x))$ or the first ocurrence of x in $A(x).\exists x.B(x)$. A variable that is both bound and trapped is non-active and fully analogous to the bound variables of classical logic.

In the dynamic world there are really two candidates for correspondence to the classical notion of *sentence* in Predicate Logic (rather like the classical concept of mass divides into two in Relativity Theory): formulas in which all variables are bound and formulas in which all variables are both bound and trapped. On our view sentences in the first sense are what describes states. Note that in contrast to sentences in Predicate Logic and to sentences in the second sense in DPL/DRT sentences in the first sense have more interesting meanings than just a truthvalue w.r.t. the given Model.

Our examples purported to illustrate the notion of state focussing on anaphoric phenomena. These are not the only relevant kind of phenomena. The processing of syntactic structure can be treated in an analogous way (at least for the admittedly modest fragments I have been considering). The simplest kind of model in this direction is what one gets when abstracting from what is between the brackets, i.o.w. when one just considers strings of brackets. Here a state is any string that has survived the bracket-test, i.e. any string where the bracket count, counting "(" as +1 and ")" as −1, has not sunk below

0. So these strings don't 'ask' for "("'s at the beginning. This is saturation towards the past. We don't ask similar saturation towards the future. The *OTAT*-principle simply tells us that when the bracket count has sunk below 0, nothing that comes after will set it right. The *SSC*'s will be extensively discussed in section 12.6.4 of this chapter.

12.2.4   The Silly Side of (Our Version of) Update Semantics

As we set it up in section 12.2.2 as soon as the output of a candidate update is not a state, the result is undefined. This seems to be definitely unrealistic. If I hear a fragment of conversation *He was smiling*, not knowing whom they are talking about, it would be simply ridiculous to 'become' undefined. For one thing it would provide people with overly simple ways of getting other people out of the way. The realistic way of handling the fragment is to set the problem of interpreting *he* aside as something to be dealt with later.

My hunch is that one should first get the silly model of updating straight before building more realistic models to describe how we actually handle semantically incomplete information. One hopeful sign is that the silly model more or less automatically leads to the notion of *partial state*. Perhaps *setting aside the problem of interpreting something* can be described as going into a partial state. (Still even partial states do not give us *error recovery*: the *OTAT*-principle blocks this.)

12.2.5   Information Orderings

Till now we have just been thinking about the merging behaviour of meanings or information contents. But definitely the picture is incomplete if there are not ways to compare information contents. We will handle this problem by assuming that our meanings come with an ordering: the information ordering.

We write our information orderings in the Heyting Algebra style: so more informative is smaller. The top is the least informative item, the bottom the most informative one (in most situations the bottom is even over-informative.)

There are in fact two kinds of information ordering. The first one is the *synchronic* information ordering. For example I have two pieces of paper in my pocket. One states *Jan is wearing something new*, the other *Jan is wearing a new hat*. Evidently the first piece of paper is less informative than the second one. Whatever information state someone is in, being offered the second piece will leave her at least as informed as being offered the first. So we compare the effects of the pieces of paper when offered at the same time to the same person in different possible situations.

The second ordering is the *diachronic* ordering. Consider *Genever is a wonderful beverage. Not only the Dutch are fond of it.* Now the information content of both *Genever is a wonderful beverage* and of *Not only the Dutch are fond of it* are part of the information content of *Genever is a wonderful beverage. Not only the Dutch are fond of it.*

But they are part by virtue of being brought into the whole via the process of consecutive presentation. Synchronic comparison of e.g. *Genever is a wonderful beverage. Not only the Dutch are fond of it* and *Not only the Dutch are fond of it* is a rather pointless exercise.

Both in the case of the synchronic ordering and of the diachronic ordering we may wish to distinguish ways in which one item is more informative than another one. This leads us to studying labeled orderings or categories, rather than ordinary orderings.

In this chapter we will only study the synchronic unlabeled ordering. We will assume that the synchronic ordering together with the merger gives rise to the rich structure of a residuation lattice. This assumption is unfortunately not based an informally rigorous analysis, but just on the fact that some important examples satisfy it. So in a later stage of research we may have to retrace our steps.

Our basic structure is a *reduced residuation lattice* $\mathcal{A} = \langle A, \vee, \wedge, \cdot, 1, \rightarrow, \leftarrow \rangle$. Define:

$$a \leq b :\Longleftrightarrow a \vee b = b.$$

Let $\mathcal{A}$ satisfy:

- $\langle A, \vee, \wedge \rangle$ is a lattice, where we do not assume the top and the bottom;
- $\langle A, \cdot, 1 \rangle$ is a monoid;
- $a \cdot b \leq c \leftrightarrow a \leq c \leftarrow b \leftrightarrow b \leq a \rightarrow c.$

$\leftarrow$ is *left residuation* or *post-implication*. $\rightarrow$ is *right residuation* or *pre-implication*. We will consider the 'real order' of the arguments of post-implication to be opposite to the displayed one.

We left out the top and bottom just for temporary convenience: we could have left them, but that would make some formulations later on a bit heavy.

There are two intuitions about the synchronic ordering. The first takes presuppositions to be informative, the second takes presuppositions to be anti-informative. So according to the second intuition the more information a content presupposes the less informative it is. As we will see later on these two intuitions correspond respectively to viewing undefined as *deadlock* or *error* and to viewing undefined as *not sufficient*.

In this chapter the second intuition will be our choice. So according to us a presupposition will be on a negative place. (Clearly it could turn out that the right approach is to keep both options at the same time. This could lead to a treatment a bit like *bilattices*.)

Our choice leads immediately to a pleasant definition of the set of states in terms of the algebra. Remember that 1 is the tabula rasa mental state: it asks for nothing, it contains nothing. The items that are more informative than 1 are precisely the ones that presuppose less than 1 and contain at least as much (static) information as 1, i.e., precisely the ones that presuppose nothing, i.e. precisely the states. So we take:

$$S := \{ a \in A \mid a \leq 1 \}.$$

a, b, c, a', b', c', ... will range over A and s, s', u, u', ... will range over S. Note that b has the *OTAT*-property precisely if for all a: $a \cdot b \leq 1 \implies a \leq 1$, i.o.w. precisely if $(1 \leftarrow b) \leq 1$. 0 (if present) has the *OTAT*-property just in case $1 = \top$.

Let's consider information orderings in terms of update functions. An update function is a partial function $F \colon S \to S$. Given an information ordering on S we can define two induced orderings, corresponding to the options we just discussed, on the update functions:

$$F \leq_1 G :\Longleftrightarrow \forall s\,(sF{\Downarrow} \implies sG{\Downarrow} \text{ and } sF \leq sG),$$

$$F \leq_2 G :\Longleftrightarrow \forall s\,(sG{\Downarrow} \implies sF{\Downarrow} \text{ and } sF \leq sG).$$

Clearly these are partial orderings.

Suppose for the moment that we would like to expand the ordering on the states with a new element 'undefined' or $\uparrow$ in such a way that $\leq_1$, respectively $\leq_2$, becomes the pointwise induced ordering. Consider the $\leq_1$-case. Noting e.g. that the nowhere defined function is the bottom, the unique way of achieving this is making $\uparrow$ a new bottom. Similarly in the $\leq_2$-case $\uparrow$ should be made a new top. In the $\leq_1$-case $\uparrow$ is even *more* informative than 'overdetermined' or 'false'. One way of understanding this is to view $\uparrow$ as an error state or a deadlock. In the $\leq_2$-case $\uparrow$ is even *less* informative than 'tabula rasa' or 'true'. One way of understanding this is to view $\uparrow$ as the not-sufficient-state, something that strives to be a state, but needs something extra to be that. (It would be somewhat misleading to say that $\uparrow$ is open-ended, since the insufficiency is more naturally thought of as being 'on the side of the past'.)

In this chapter we will study a specific set of update functions: the ones that update by merging with a fixed element a of our residuation lattice and whose domain is given as the set of all s below or equal to a fixed $a' \leq 1$. The idea is that a' represents a condition on the states: states below a' carry sufficient information to get access to the updating element a. (Note that the word *condition* is used here from the external point of view of the theoretician, not from the internal point of view of the framework.)

It may seem somewhat strange that no *intrinsic connection* is demanded between a' and a, but this can be understood by realizing that the update functions are supposed to be semantical objects. The coming together of presupposition state and update action is imported from the level of language use. If for example someone tells me *The present king of France is bald* the information contained in this sentence can only be processed by those having states providing a present king of France. The update simply has the form: *x is bald*. No intrinsic connection is called for between kings of France and baldness.

We pick up the theme of updates again in section 12.4.

EXCURSION 12.2.5.1 (VALIDITY AND IMPLICATION)   One of the major problems of the DPL/DRT approach is to gain an algebraic understanding of validity and implication.

To give the reader some feeling for the problem let's briefly consider the problem in the case of Groenendijk & Stokhof's DPL.

DPL-meanings are relations between assignments. The merger simply becomes relation composition. We need also dynamic implication $\hookrightarrow$, where

$$f(R \hookrightarrow S)g :\Longleftrightarrow f = g \text{ and } \forall h\,(fRh \Longrightarrow \exists i\,hSi).$$

Given a classical model $\mathcal{M}$ we may define:

$$f \,\|\exists x\|\, g :\Longleftrightarrow \text{ for all variables } y \text{ different from } x: \; yf = yg,$$
$$f \,\|Px\|\, g :\Longleftrightarrow f = g \text{ and } xf \in \|P\|,$$
$$\|\bot\| := \emptyset$$
$$\|\varphi.\psi\| := \|\varphi\| \circ \|\psi\|$$
$$\|\varphi \to \psi\| := \|\varphi\| \hookrightarrow \|\psi\|$$
$$\varphi_0, \ldots, \varphi_{n-1} \models_{\mathcal{M},f} \psi :\Longleftrightarrow \forall g\,(f\,\|\varphi_0\| \circ \cdots \circ \|\varphi_0\|\,g \Longrightarrow \exists h\,g\,\|\psi\|\,h)$$

$\forall x\,(\varphi)$ can e.g. be considered as an abbreviation of: $(\exists x \to \varphi)$.

At first sight some very basic progress is made here: we have before us a definition of (a form of) Predicate Logic that is a genuine special case of the corresponding version of propositional logic. The existential quantifier is just an atom, linked with the rest of the text by the propositional connectives. Granted, this is true. On the other hand, however, the propositional 'algebra' we have here is definitely unattractive. (i) As far as I know we have no axiomatization of the logic of $\circ$ and $\hookrightarrow$ for the binary relations over an arbitrary domain. (ii) $\hookrightarrow$ not only handles 'negative place' but also throws away internal values assigned to variables. (iii) $\hookrightarrow$ is non-transitive. (As is illustrated by van Benthem's example: *Everyone who has a house, has a garden. Everyone who has a garden sprinkles it. But not: Everyone who has a house sprinkles it.*) (iv) Repetition of $\exists x$ with the same variable, is an obnoxious bug in the system, since it has the effect of throwing away all information about the values of the first occurrence of x.

Similar problems haunt also other related semantics like the one of DRT.

Johan van Benthem suggests to define $\hookrightarrow$ in terms of other more basic operations. This suggestion is surely on the right track, but just as surely not every definition can count as success. E.g. in the residuation lattice of relations over a given domain extended with the converse-operation $^\vee$, the operation $\hookrightarrow$ can be defined as follows: $R \hookrightarrow S := 1 \wedge ((\top \circ S^\vee) \leftarrow R)$.

Proof

$$u(1 \wedge ((\top \circ S^\vee) \leftarrow R))v \Longleftrightarrow u = v \text{ and } \forall w\,(vRw \Longrightarrow u(\top \circ S^\vee)w)$$
$$\Longleftrightarrow u = v \text{ and } \forall w\,(vRw \Longrightarrow \exists z\,u\top z S^\vee w)$$

$$\Longleftrightarrow u = v \text{ and } \forall w \, (vRw \Longrightarrow \exists z \, zS^\vee w)$$
$$\Longleftrightarrow u = v \text{ and } \forall w \, (vRw \Longrightarrow \exists z \, wSz) \qquad \blacksquare$$

I submit, however, that this definition is too ad hoc to be enlightening. The problem is what Henk Barendregt calls, slightly adapting a Zen usage, a Koan. This means that what the problem is only becomes fully clear when we see the solution.

Open problem Can $\hookrightarrow$ be defined in the residuation lattice of relations (without using $^\vee$)?

(Lysbeth Zeinstra in her master's thesis defines (in a slightly different setting) $\hookrightarrow$ from a binary connective "*so*". In this approach the 'trapping' of the variables in implications is effected by explicit 'downdates'. Still most of the problems of $\hookrightarrow$ also plague "*so*".)

In the present chapter we will touch on the problems surrounding validity and implication only in passing.

12.3   Some Elementary Facts Concerning Residuation Lattices

In this section some simple constructions in residuation lattices are described. More information on residuation algebras and action algebras can be found in Pratt [15] and Kozen [12].

Residuation lattices have obvious connections to category theory and linear logic (for the last see e.g. Abrusci [1]). Other close relatives are the bilattices due to Ginsberg (see e.g. Fitting [9]; in fact the construction described in that paper bears some similarity to our work in section 12.4).

Consider an a reduced residuation lattice $\mathcal{A} = \langle A, \vee, \wedge, \cdot, 1, \rightarrow, \leftarrow \rangle$ (as introduced in 12.2.5). Define:

$a \le b :\Longleftrightarrow a \vee b = b;$

$b \mathrel{{}_1{\leftarrow}} a := (b \leftarrow a) \wedge 1;$

$\qquad S := \{ a \in A \mid a \le 1 \}.$

$a, b, c, a', b', c', \dots$ will range over A and $s, s', u, u', \dots$ will range over S.

For completeness we state some principles valid in a (reduced) residuation lattice without proof. We only state principles for $\leftarrow$, but of course the corresponding ones for $\rightarrow$ also hold. The statements involving 0 and $\top$ only apply, when 0 is present.

$$x \cdot (y \vee z) = x \cdot y \vee x \cdot z,$$
$$(x \vee y) \cdot z = x \cdot z \vee y \cdot z,$$
$$0 \cdot x = x \cdot 0 = 0,$$

$$x \leftarrow (y \vee z) = (x \leftarrow y) \wedge (x \leftarrow z),$$
$$(x \wedge y) \leftarrow z = (x \leftarrow z) \wedge (y \leftarrow z),$$
$$x \leftarrow y \cdot z = (x \leftarrow z) \leftarrow y,$$
$$x \leftarrow 0 = \top \leftarrow x = \top,$$
$$x \leftarrow 1 = x,$$
$$(x \leftarrow y) \cdot y \leq x,$$
$$(x \leftarrow y) \cdot (y \leftarrow z) \leq (x \leftarrow z).$$

We will amply use the following proof-generated property:

$$s' \cdot a \leq s' \cdot b \text{ and } s \leq s' \Longrightarrow s \cdot a \leq s \cdot b. \tag{Ω}$$

Ω is equivalent to:

$$s \leq s' \Longrightarrow (s' \rightarrow s' \cdot b) \leq (s \rightarrow s \cdot b).$$

So Ω says in a sense that in case of repetitions of states on the left before and after an inequality the occurrence on the negative place is the one that weights heavier.

We give an example of a residuation lattice not satisfying Ω. Let $A := \{0, \mathbf{u}, \mathbf{v}, 1\}$, where $0 < \mathbf{u} < \mathbf{v} < 1$, $\mathbf{u} \cdot \mathbf{u} = \mathbf{u} \cdot \mathbf{v} = \mathbf{v} \cdot \mathbf{u} = 0$ and $\mathbf{v} \cdot \mathbf{v} = \mathbf{v}$. It is easy to verify that this determines a residuation lattice (even an action lattice). Ω fails because $\mathbf{v} \cdot 1 \leq \mathbf{v} \cdot \mathbf{v}$, but $\mathbf{u} \cdot 1 \not\leq \mathbf{u} \cdot \mathbf{v}$.

$\cdot$	0	u	v	1		$\rightarrow$	0	u	v	1
0	0	0	0	0		0	1	1	1	1
u	0	0	0	u		u	v	1	1	1
v	0	0	v	v		v	u	u	1	1
1	0	u	v	1		1	0	u	v	1

Table 12.1
Truthtables for our example.

There is a property that is somewhat more natural (but prima facie stronger) than Ω, a strengthening of Modus Ponens in case the antecedent is a state: SMP is the property: $(a \;_1\!\!\leftarrow s) \cdot s = s \wedge a$. SMP_1 is SMP for $a \leq 1$.

FACT 12.1   (i) Ω follows from SMP_1; (ii) SMP follows from SMP_1.

Proof (i) Suppose SMP_1 and $s \cdot a \leq s \cdot b$ and $u \leq s$. Then $(u \,_1\!\!\leftarrow s) \cdot s \cdot a \leq (u \,_1\!\!\leftarrow s) \cdot s \cdot b$ and hence $u \cdot a \leq u \cdot b$. (ii) Suppose SMP_1. Then:

$$(a \,_1\!\!\leftarrow s) \cdot s = ((s \wedge a) \,_1\!\!\leftarrow s) \cdot s = s \wedge a. \qquad \blacksquare$$

Any set S' that (i) satisfies Ω (in the sense that we let the variables s and s' in the statement of Ω range over S'), that (ii) has a maximum m, (iii) is closed under $\cdot$, (iv) is downwards closed and (v) contains 1, is equal to S. In other words: $m = 1$. Simply note that: $m \cdot m \leq m = m \cdot 1$. Apply Ω with $s := m$ and $s' := 1$, using $m \cdot m = m \cdot 1$ and $1 \leq m$. We find: $m = 1 \cdot m \leq 1 \cdot 1 = 1$.

12.3.1   The Algebra $\mathcal{A}_m$

Suppose $1 \leq m$ and $m \cdot m \leq m$. $\mathcal{A}_m$ is the (reduced) residuation lattice obtained by restricting the domain to $A_m := \{\, a \mid a \leq m \,\}$ and by taking $\vee, \wedge, \cdot, 1$ as before (it is easily seen that this can be done) and by taking as residuations $\rightarrow_m$ and $_m\!\!\leftarrow$, where:

$$a \rightarrow_m b := (a \rightarrow b) \wedge m, \quad \text{and} \quad b \,_m\!\!\leftarrow a := (b \leftarrow a) \wedge m.$$

It is easily verified that the resulting algebra is as desired.

The specific example we will meet later is of course $\mathcal{A}_1$. As is easily seen the construction under consideration also preserves Kleene's $*$, so if we start with an action lattice we get a new action lattice.

12.3.2   The Relation $\leq_d$

We collect some facts about the ordering $\leq_d$, which will be useful later.

Define: $a \leq_d b :\Longleftrightarrow d \cdot a \leq d \cdot b$ and $a =_d b :\Longleftrightarrow d \cdot a = d \cdot b$. Clearly $\leq_d$ is a preordering with induced equivalence relation $=_d$. Below we will now and then conveniently confuse $\leq_d$ with its induced ordering on the $=_d$ equivalence classes. Note that $a \leq b \Longrightarrow a \leq_d b$. Define $N_d(b) := (d \rightarrow d \cdot b)$. As is well known N_d is a closure operation. We have:

$$a \leq_d b \Longrightarrow a \leq N_d(b)$$
$$\Longrightarrow d \cdot a \leq d \cdot N_d(b) \leq d \cdot b$$
$$\Longrightarrow a \leq_d b.$$

So $N_d(b)$ is the maximal element of the $=_d$-equivalence class of b.

$a \vee b$ is the $\leq_d$-suprememum of a and b, since:

$$a \vee b \leq_d c \Longleftrightarrow d \cdot a \vee d \cdot b \leq d \cdot c$$
$$\Longleftrightarrow a \leq_d c \text{ and } b \leq_d c.$$

$N_d(a) \wedge N_d(b)$ is the $\leq_d$-infimum of a and b, since:

$$c \leq_d N_d(a) \wedge N_d(b) \Longrightarrow c \leq_d N_d(a) \text{ and } c \leq_d N_d(b)$$
$$\Longrightarrow c \leq N_d(a) \text{ and } c \leq N_d(b)$$
$$\Longrightarrow c \leq N_d(a) \wedge N_d(b)$$
$$\Longrightarrow c \leq_d N_d(a) \wedge N_d(b).$$

Moreover $(d \to (d \cdot c \leftarrow b))$ is a kind of post-implication for $\leq_d$ and $\cdot$:

$$a \cdot b \leq_d c \Longrightarrow d \cdot a \leq d \cdot c \leftarrow b$$
$$\Longrightarrow a \leq d \to (d \cdot c \leftarrow b)$$
$$\Longrightarrow d \cdot a \leq d \cdot (d \to (d \cdot c \leftarrow b)) \leq d \cdot c \leftarrow b$$
$$\Longrightarrow a \cdot b \leq_d c.$$

Note however that $\cdot$ is an operation modulo $=_d$ only in the first argument.

Finally in the presence of Ω we have yet another desirable property. Suppose Ω and $s \leq u \leq 1$, then:

$$c \leq_u N_s(a) \Longrightarrow c \leq_s N_s(a)$$
$$\Longrightarrow c \leq N_s(a)$$
$$\Longrightarrow c \leq_u N_s(a).$$

Ergo: $c \leq N_s(a) \leftrightarrow c \leq_u N_s(a) \leftrightarrow c \leq N_u N_s(a)$, and so $N_u N_s(a) = N_s(a)$.

12.4 Update Functions and Partial Actions

In this section we present the main construction of partial actions and prove its basic properties. The partial actions can be viewed (except for a few special elements) as update functions on states.

Fix a reduced residuation lattice $\mathcal{A}$. Consider the update functions on $\mathcal{A}$. We will consider these as ordered by $\leq_2$ of 12.2.5. We will designate $\leq_2$ simply by $\leq$. Remember that with every element a we associate its canonical update function Φ_a, given by: $s\Phi_a := s \cdot a$ if $s \cdot a \in S$, $s\Phi_a\uparrow$ otherwise. Note that:

$$s \cdot a \in S \leftrightarrow s \cdot a \leq 1 \leftrightarrow s \leq 1 \leftarrow a \leftrightarrow s \leq 1 \,{}_1\!\!\leftarrow a.$$

(Reminder: $b \,{}_1\!\!\leftarrow a =: (b \leftarrow a) \wedge 1$.) Thus $1 \,{}_1\!\!\leftarrow a$ is the canonical presupposition for updating with a. We define: $pre(a) := 1 \,{}_1\!\!\leftarrow a$. The class of canonical update functions is not always a good class: e.g. if $OTAT$ fails for $\mathcal{A}$ it may not be closed under composition.

We will study a somewhat larger class of update functions. These will be given by a presupposition state and an update action. Such updates can be considered as partial actions. (As we will see in section 12.4.1 this is slightly misleading, for even if there is a 'canonical embedding' of $\mathcal{A}$ into the algebra given by these updates, $\leq$ and function composition, this embedding need not be a morphism of reduced residuation lattices.)

Consider a pair $\alpha := \langle a', a \rangle$ where $a' \in S$ and $a' \cdot a \in S$ (or: $a' \leq pre(a)$). Let $\Psi_\alpha : S \to S$ be given by: $s\Psi_\alpha := s \cdot a$ if $s \leq a'$, $:= \uparrow$ otherwise. Note that $s \cdot a \leq a' \cdot a \leq 1$, so $s \cdot a \in S$.

Note that $\Phi_a = \Psi_{\langle pre(a), a \rangle}$.

Let $U := \{ \langle a', a \rangle \mid a' \in S, a' \cdot a \in S \}$. $X := \{ \Psi_\alpha : S \to S \mid \alpha \in U \}$. We show that X is closed under $\circ$.

FACT 12.2 X is closed under $\circ$.

Proof

$$s\Psi_{\langle a', a \rangle} \circ \Psi_{\langle b', b \rangle} \Downarrow \iff s \leq a' \text{ and } s \cdot a \leq b'$$
$$\iff s \leq a' \wedge (b' \leftarrow a).$$

Moreover if $s\Psi_{\langle a', a \rangle} \circ \Psi_{\langle b', b \rangle} \Downarrow$, then $s\Psi_{\langle a', a \rangle} \circ \Psi_{\langle b', b \rangle} = s \cdot a \cdot b$. So:

$$\Psi_{\langle a', a \rangle} \circ \Psi_{\langle b', b \rangle} = \Psi_{\langle a' \wedge (b' \leftarrow a), a \cdot b \rangle}. \qquad \blacksquare$$

It will be convenient to talk about the pairs $\langle a', a \rangle$ instead of about the corresponding update functions. To do this we need to know the induced merger and the induced preorder and corresponding equivalence relation on the pairs. Define:

$$\langle a', a \rangle \cdot \langle b', b \rangle := \langle a' \wedge (b' \leftarrow a), a \cdot b \rangle.$$

The proof of 12.2 gives us:

FACT 12.3 $\Psi_{\langle a', a \rangle} \circ \Psi_{\langle b', b \rangle} = \Psi_{\langle a', a \rangle \cdot \langle b', b \rangle}.$

To get a nice characterization of our preordering and equivalence relation we stipulate as our STANDING ASSUMPTION that $\mathcal{A}$ satisfies Ω!

Define:

$$\langle a', a \rangle \leq \langle b', b \rangle :\iff b' \leq a' \text{ and } b' \cdot a \leq b' \cdot b,$$
$$\langle a', a \rangle \equiv \langle b', b \rangle :\iff b' = a' \text{ and } b' \cdot a = b' \cdot b.$$

We have:

FACT 12.4

$$\langle a', a \rangle \le \langle b', b \rangle \iff \Psi_{\langle a', a \rangle} \le \Psi_{\langle b', b \rangle},$$
$$\langle a', a \rangle \equiv \langle b', b \rangle \iff \Psi_{\langle a', a \rangle} = \Psi_{\langle b', b \rangle}.$$

Proof

$$\Psi_{\langle a', a \rangle} \le \Psi_{\langle b', b \rangle} \iff \forall s \le b' \, (s \le a' \text{ and } s \cdot a \le s \cdot b)$$
$$\iff b' \le a' \text{ and } b' \cdot a \le b' \cdot b.$$

And:

$$\Psi_{\langle a', a \rangle} = \Psi_{\langle b', b \rangle} \iff \langle a', a \rangle \le \langle b', b \rangle \text{ and } \langle b', b \rangle \le \langle a', a \rangle$$
$$\iff a' = b' \text{ and } b' \cdot a = b' \cdot b. \qquad \blacksquare$$

We work with the representatives $\langle a', a \rangle$. But, of course, the real objects we are considering are the equivalence classes of $\equiv$ and via these the elements of X!

Let $Y := X \cup \{0\}$ if $\mathcal{A}$ has a bottom, $:= X \cup \{0, \top\}$ otherwise. We extend $\le$ on Y by making 0 the bottom and (in the second case) $\top$ the top. Define $F \cdot G := F \circ G$, $F \cdot 0 = 0 \cdot F = 0$, $F \cdot \top = \top \cdot G = \top$, $\top \cdot 0 = 0 \cdot \top = 0$. (Reader, please don't worry about the top and bottom at this point. Why they are added in this specific way will become clear in the proof of 12.2)

Let $\mathcal{U}_0(\mathcal{A}) := \langle Y, \le, \cdot \rangle$. The main result of this section is:

THEOREM 12.2   Let $\mathcal{A}$ be a reduced residuation lattice satisfying Ω, then:

i.   There is a unique residuation lattice $\mathcal{U}(\mathcal{A}) = \langle Y, \vee, \wedge, 0, 1, \cdot, \rightarrow, \leftarrow \rangle$, extending $\mathcal{U}_0(\mathcal{A})$, i.e. there is a unique residuation lattice $\mathcal{U}(\mathcal{A}) = \langle Y, \vee, \wedge, 0, 1, \cdot, \rightarrow, \leftarrow \rangle$, such that the order based on $\vee$ is the ordering $\le$ of $\mathcal{U}_0(\mathcal{A})$ and $\cdot$ is the same in $\mathcal{U}(\mathcal{A})$ and $\mathcal{U}_0(\mathcal{A})$.

ii.   If $\mathcal{A}$ is an action lattice then $\mathcal{U}_0(\mathcal{A})$ can be extended in a unique way to an action lattice $\mathcal{U}^*(\mathcal{A})$.

We will be forced to introduce 0 and $\top$ even if we were only aiming to find a reduced residuation lattice.

Proof   It is well known that if a structure of the form $\mathcal{U}_0(\mathcal{A})$ can be extended to a residuation (action) lattice (in the sense given above), then such an extension is unique. So the only thing we need to do to prove the theorem is to 'compute' the desired operations.

We first treat $\wedge$ and $\vee$ for the pairs (the treatment for 0 and $\top$ being selfevident). Define:

$$\langle a', a \rangle \vee \langle b', b \rangle := \langle a' \wedge b', a \vee b \rangle$$
$$\langle a', a \rangle \wedge \langle b', b \rangle := \langle a' \vee b', N_{a'}(a) \wedge N_{b'}(b) \rangle.$$

It is easily seen that the values are in U.

$\vee$ is the supremum w.r.t. $\leq$ and $\wedge$ is the infimum: let $\alpha = \langle a', a \rangle$, $\beta = \langle b', b \rangle$, $\gamma = \langle c', c \rangle$, since:

$$\alpha \vee \beta \leq \gamma \Longleftrightarrow c' \leq a' \wedge b' \text{ and } a \vee b \leq_{c'} c$$
$$\Longleftrightarrow c' \leq a', \ c' \leq b', \ a \leq_{c'} c, \ b \leq_{c'} c$$
$$\Longleftrightarrow \alpha \leq \gamma \text{ and } \beta \leq \gamma;$$
$$\gamma \leq \alpha \wedge \beta \Longleftrightarrow a' \vee b' \leq c' \text{ and } c \leq_{a' \vee b'} N_{a'}(a) \wedge N_{b'}(b)$$
$$\Longleftrightarrow a' \leq c', \ b' \leq c', \ c \leq_{a' \vee b'} N_{a' \vee b'} N_{a'}(a) \wedge N_{a' \vee b'} N_{b'}(b)$$
$$\Longleftrightarrow a' \leq c', \ b' \leq c', \ c \leq N_{a' \vee b'} N_{a'}(a) \text{ and } c \leq N_{a' \vee b'} N_{b'}(b)$$
$$\Longleftrightarrow a' \leq c', \ b' \leq c', \ c \leq N_{a'}(a) \text{ and } c \leq N_{b'}(b)$$
$$\Longleftrightarrow \langle c', c \rangle \leq \langle a', a \rangle \text{ and } \langle c', c \rangle \leq \langle b', b \rangle.$$

(The reader is referred to 12.3.2 for the relevant facts on N which are used here.)

What about top and bottom in the new algebra? Since we always add a fresh bottom, the bottom can give no problems. If $\mathcal{A}$ has no bottom, we add a new top. So again no problems. Suppose finally $\mathcal{A}$ has a bottom 0. Clearly $0 \in S$, and so $\langle 0, 0 \rangle \in U$. We get: $\langle a', a \rangle \leq \langle 0, 0 \rangle \leftrightarrow 0 \leq a'$ and $0 \cdot a \leq 0 \cdot 0$. So $\langle 0, 0 \rangle$ becomes the top of our new algebra. We will in the last case, perhaps confusingly, also designate $\langle 0, 0 \rangle$ by: $\top$.

EXCURSION 12.4.0.1   In case 0 is present in $\mathcal{A}$, we also have:

$$\langle 1, 0 \rangle \leq \langle a', a \rangle \leftrightarrow a' \leq a \text{ and } a' \cdot 0 \leq a' \cdot a.$$

So $\langle 1, 0 \rangle$ is the least of the pairs. We have:

$$\langle 1, 0 \rangle \cdot \langle b', b \rangle = \langle 1 \wedge (b' \leftarrow 0), 0 \cdot b \rangle = \langle 1, 0 \rangle$$
$$\langle a', a \rangle \cdot \langle 1, 0 \rangle = \langle a' \wedge (1 \leftarrow a), a \cdot 0 \rangle = \langle a', 0 \rangle \qquad \text{(remember that } a' \cdot a \leq 1\text{).}$$

The last equation tells us that $\langle 1, 0 \rangle$ is not an annihilator for $\cdot$ and that thus $\langle 1, 0 \rangle$ can not play the role of the 0 of a residuation algebra. Hence adding the new bottom is really necessary.

We turn to the unit element of the merger. Let $1 := \langle 1, 1 \rangle$. We have

$$1 \cdot \langle b', b \rangle = \langle 1 \wedge (b' \leftarrow 1), 1 \cdot b \rangle = \langle 1 \wedge b', b \rangle = \langle b', b \rangle$$

and

$$\langle a', a \rangle \cdot 1 = \langle a' \wedge (1 \leftarrow a), a \cdot 1 \rangle = \langle a', a \rangle,$$

since $a' \cdot a \leq 1$ and hence $a' \leq 1 \leftarrow a$.

EXCURSION 12.4.0.2   In case our algebra has a bottom 0, $\langle 0, 0 \rangle$ has the role of top. We have:

$$\top \cdot \langle b', b \rangle = \langle 0, 0 \rangle \cdot \langle b', b \rangle = \langle 0 \wedge (b' \leftarrow 0), 0 \cdot b \rangle = \langle 0, 0 \rangle = \top,$$
$$\langle a', a \rangle \cdot \top = \langle a', a \rangle \cdot \langle 0, 0 \rangle = \langle a' \wedge (0 \leftarrow a), a \cdot 0 \rangle.$$

So the non-added top, doesn't quite behave like the added top!

We proceed with the computation of the residuations. First the "basic equation":

$$\langle a', a \rangle \cdot \langle b', b \rangle \leq \langle c', c \rangle \Longleftrightarrow \langle a' \wedge (b' \leftarrow a), a \cdot b \rangle \leq \langle c', c \rangle$$
$$\Longleftrightarrow c' \leq a' \wedge (b' \leftarrow a) \text{ and } c' \cdot a \cdot b \leq c' \cdot c.$$

We 'solve' $\langle a', a \rangle$ from the rhs.:

$$\langle a', a \rangle \cdot \langle b', b \rangle \leq \langle c', c \rangle$$
$$\leftrightarrow c' \leq a' \text{ and } c' \cdot a \leq b' \text{ and } c' \cdot a \cdot b \leq c' \cdot c$$
$$\leftrightarrow c' \leq a' \text{ and } c' \cdot a \leq c' \cdot ((c' \to b') \wedge (c' \to (c' \cdot c \leftarrow b))).$$

We prove the non trivial part of the second equivalence: let P be:

$$c' \cdot a \leq c' \cdot ((c' \to b') \wedge (c' \to (c' \cdot c \leftarrow b))).$$

"$\Longrightarrow$" Suppose $c' \cdot a \leq b'$ and $c' \cdot a \cdot b \leq c' \cdot c$, then $a \leq c' \to b'$ and $a \leq c' \to (c' \cdot c \leftarrow b)$. Ergo: $a \leq (c' \to b') \wedge (c' \to (c' \cdot c \leftarrow b))$ and hence P.
  "$\Longleftarrow$" Suppose P. Then $c' \cdot a \leq c' \cdot (c' \to b') \leq b'$ and

$$c' \cdot a \leq c' \cdot (c' \to (c' \cdot c \leftarrow b)) \leq c' \cdot c \leftarrow b,$$

hence $c' \cdot a \cdot b \leq c' \cdot c$.
  Define:

$$\langle c', c \rangle \leftarrow \langle b', b \rangle := \langle c', (c' \to b') \wedge (c' \to (c' \cdot c \leftarrow b)) \rangle.$$

Clearly:

$$\langle a', a \rangle \leq \langle c', c \rangle \leftarrow \langle b', b \rangle$$
$$\leftrightarrow c' \leq a' \text{ and } c' \cdot a \leq c' \cdot ((c' \to b') \wedge (c' \to (c' \cdot c \leftarrow b))).$$

We 'solve' $\langle b', b \rangle$:

$$\langle a', a \rangle \cdot \langle b', b \rangle \leq \langle c', c \rangle$$
$$\leftrightarrow c' \leq a' \text{ and } c' \cdot a \leq b' \text{ and } c' \cdot a \cdot b \leq c' \cdot c$$
$$\leftrightarrow c' \leq a' \text{ and } c' \cdot a \leq b' \text{ and } c' \cdot a \cdot b \leq c' \cdot a \cdot (c' \cdot a \to c' \cdot c).$$

Here we meet a problem: what to do with the clause $c' \leq a'$, which is independent of b' and b? The solution is use our new bottom 0. Set:

$$\langle a', a \rangle \to \langle c', c \rangle := \langle c' \cdot a, c' \cdot a \to c' \cdot c \rangle \text{ if } c' \leq a', \; := 0 \text{ otherwise.}$$

In case $c' \leq a$ we have:

$$\langle b', b \rangle \leq \langle a, a \rangle \to \langle c', c \rangle \leftrightarrow c' \cdot a \leq b' \text{ and } c' \cdot a \cdot b \leq c' \cdot a \cdot (c' \cdot a \to c' \cdot c).$$

In case not $c' \leq a'$:

$$\langle b', b \rangle \leq \langle a', a \rangle \to \langle c', c \rangle \leftrightarrow \langle b', b \rangle \leq 0.$$

Combining these we find the desired:

$$\langle b', b \rangle \leq \langle a', a \rangle \to \langle c', c \rangle$$
$$\leftrightarrow c' \leq a' \text{ and } c' \cdot a \leq b' \text{ and } c' \cdot a \cdot b \leq c' \cdot a \cdot (c' \cdot a \to c' \cdot c).$$

How is 0 going to behave w.r.t. $\leftarrow$? It is easy to see that we should have for $\alpha \in U \cup \{0, \top\}$ (where the top may be either added or constructed):

$$\alpha \leftarrow 0 := \top; \qquad 0 \leftarrow \alpha := 0 \text{ for } \alpha \neq 0.$$

If we have a new top, we still must define its interaction with the residuations. A simple computation shows that we must set:

$$\alpha \leftarrow \top := 0 \text{ for } \alpha \neq \top; \qquad \top \leftarrow \alpha := \top.$$

SUMMARY 12.4.0.3 We restrict ourselves to the case where a new top is added. We give the connectives by specifying the corresponding operations on the representing pairs $\langle a', a \rangle$. The values of the operations on arguments involving $\{0, \top\}$ are specified in truthtables.

$$\langle a', a \rangle \vee \langle b', b \rangle := \langle a' \wedge b', a \vee b \rangle,$$
$$\langle a', a \rangle \wedge \langle b', b \rangle := \langle a' \vee b', (a' \to a' \cdot a) \wedge (b' \to b' \cdot b) \rangle$$
$$= \langle a' \vee b', N_{a'}(a) \wedge N_{b'}(b) \rangle,$$

$$\langle a', a \rangle \cdot \langle b', b \rangle := \langle a' \wedge (b' \leftarrow a), a \cdot b \rangle,$$
$$\langle b', b \rangle \leftarrow \langle a', a \rangle := \langle b', (b' \to a') \wedge (b' \to (b' \cdot b \leftarrow a)) \rangle$$
$$= \langle b', b' \to (a' \wedge (b' \cdot b \leftarrow a)) \rangle,$$
$$\langle a', a \rangle \to \langle b', b \rangle := \langle b' \cdot a, b' \cdot a \to b' \cdot b \rangle \text{ if } b' \leq a', \ := 0 \text{ otherwise.}$$

In table 12.2 we give the promised truthtables.

·	0	1	α	$\top$
0	0	0	0	0
1	0	1	α	$\top$
α	0	α		$\top$
$\top$	0	$\top$	$\top$	$\top$

$\to$	0	1	α	$\top$
0	$\top$	$\top$	$\top$	$\top$
1	0	1	α	$\top$
α	0			$\top$
$\top$	0	0	0	$\top$

Table 12.2
Truthtables for 0 and $\top$.

The truthtables for $\to$ and $\leftarrow$ happen to be the same, so we only give $\to$. α represents the 'generic' element.

QUESTION 12.1   Does our new algebra still satisfy Ω?

Finally suppose $\mathcal{A}$ is an action lattice. We give Kleene's * in $\mathcal{U}(\mathcal{A})$. Define:

$$\langle a', a \rangle^* := \langle a' \ _1\!\leftarrow a^*, a^* \rangle,$$
$$0^* := 1,$$
$$\top^* := \top \qquad \text{(in case } \top \text{ is added).}$$

Note that $(a' \ _1\!\leftarrow a^*) \cdot a^* \leq a' \leq 1$. In case $\mathcal{A}$ has bottom 0, we have:

$$\top^* = \langle 0, 0 \rangle^* = \langle 0 \ _1\!\leftarrow 0^*, 0^* \rangle = \langle 0 \ _1\!\leftarrow 1, 1 \rangle = \langle 0, 1 \rangle \equiv \top.$$

We check that * has the desired properties on the pairs, leaving the other elements to the industrious reader:

$$\langle 1, 1 \rangle \leq \langle a', a \rangle^*, \quad \text{since } a' \ _1\!\leftarrow a^* \leq 1 \text{ and } (a' \ _1\!\leftarrow a^*) \cdot 1 \leq (a' \ _1\!\leftarrow a^*) \cdot a^*,$$
$$\text{since } 1 \leq a^*.$$

And:

$$\langle a', a\rangle^* \cdot \langle a', a\rangle^* = \langle a' _1\!\leftarrow a^*, a^*\rangle \cdot \langle a' _1\!\leftarrow a^*, a^*\rangle$$
$$= \langle (a' _1\!\leftarrow a^*) \wedge ((a' _1\!\leftarrow a^*) \leftarrow a^*), a^* \cdot a^*\rangle$$
$$= \langle 1 \wedge (a' \leftarrow a^*) \wedge (1 \leftarrow a^*) \wedge (a' \leftarrow a^* \cdot a^*), a^* \cdot a^*\rangle$$
$$= \langle a' _1\!\leftarrow a^*, a^* \cdot a^*\rangle \quad (a' \le 1, \quad a^* \cdot a^* \le a^*)$$
$$\le \langle a', a\rangle^*.$$

Finally:

$$\langle a', a\rangle \le \langle a', a\rangle^*,$$

since $a' _1\!\leftarrow a^* \le a'$ and $(a' _1\!\leftarrow a^*) \cdot a \le (a' _1\!\leftarrow a^*) \cdot a^*$,

since $1 \le a^*$, and hence $(a' \leftarrow a^*) \le (a' \leftarrow 1) = a'$, and $a \le a^*$.

Consider any $\langle b', b\rangle$ and suppose:

$$\langle 1, 1\rangle \le \langle b', b\rangle, \quad \langle a', a\rangle \le \langle b', b\rangle \quad \text{and} \quad \langle b', b\rangle \cdot \langle b', b\rangle \le \langle b', b\rangle.$$

Then: $1 \le N_{b'}(b)$, $a \le N_{b'}(b)$ and $N_{b'}(b) \cdot N_{b'}(b) \le N_{b'}(b)$, so $a^* \le N_{b'}(b)$. Moreover $b' \le b' \wedge (b' \leftarrow b)$ and hence $b' \cdot b \le b'$. Since $a^* \le N_{b'}(b)$, we have $b' \cdot a^* \le b' \cdot b \le b' \le a'$ and so $b' \le a' _1\!\leftarrow a^*$. It follows that $\langle a', a\rangle^* \le \langle b', b\rangle$.

How does the old algebra fit into the new one? Consider any element a of $\mathcal{A}$. Consider the update function Φ_a. Remember that $pre(a) := 1 _1\!\leftarrow a$, and that $s\Phi_a$ is defined iff $s \le pre(a)$. Thus $\Phi_a = \Psi_{\langle pre(a), a\rangle}$. The natural embedding from $\mathcal{A}$ into $\mathcal{U}(\mathcal{A})$ is given by: $emb(a) := \langle pre(a), a\rangle$.

12.4.1 Special Elements in $\mathcal{A}$

The mapping $a \mapsto \Phi_a$ induces an equivalence relation on $\mathcal{A}$. Let $N(a) := N_{pre(a)}(a)$. We show that $N(a)$ is the maximal element equivalent to a. We have $a \le N(a)$ and so $pre(N(a)) \le pre(a)$. Note that $\langle pre(a), a\rangle \equiv \langle pre(a), N(a)\rangle$ and hence $pre(a) \cdot N(a) \le 1$. Thus $pre(a) \le pre(N(a))$. It follows that $pre(a) = pre(N(a))$. So $emb(N(a)) = emb(a)$. On the other hand if $emb(b) = \langle pre(a), b\rangle \equiv \langle pre(a), a\rangle$, then $b \le N(a)$.

12.4.2 Some Properties of emb

In general emb doesn't behave like a morphism. We *do* have:

$$emb(a \vee b) = \langle 1 _1\!\leftarrow (a \vee b), a \vee b\rangle$$
$$= \langle (1 _1\!\leftarrow a) \wedge (1 _1\!\leftarrow b), a \vee b\rangle$$
$$= emb(a) \vee emb(b).$$

So, a fortiori, *emb* is order preserving.
  Let $c := (1 \, _1{\leftarrow} \, a) \vee (1 \, _1{\leftarrow} \, b)$, then

$$emb(a \wedge b) = \langle 1 \, _1{\leftarrow} \, (a \wedge b), a \wedge b \rangle$$
$$\leq \langle c, N_c(a) \wedge N_c(b) \rangle$$
$$= emb(a) \wedge emb(b).$$

EXAMPLE 12.1   To see that we cannot do better consider e.g. the algebra with domain
0, 1, **a**, **b**, ⊤. The ordering relation is given by figure 12.1.

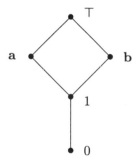

Figure 12.1
The ordering relation of our example

We stipulate that $0 \cdot x := x \cdot 0 := 0$, $1 \cdot x := x \cdot 1 := x$, for $x \neq 0$: $x \cdot \top := \top \cdot x := \top$,
$\mathbf{a} \cdot \mathbf{a} := \mathbf{a}$, $\mathbf{b} \cdot \mathbf{b} := \mathbf{b}$, $\mathbf{a} \cdot \mathbf{b} := \mathbf{b} \cdot \mathbf{a} := \top$. It is easy to check that these stipulations
determine a residuation lattice (even an action lattice) satisfying Ω. Note that $\cdot$ is
commutative and idempotent.
  We find that $emb(\mathbf{a} \wedge \mathbf{b}) = emb(1) = \langle 1, 1 \rangle = 1$, $emb(\mathbf{a}) = \langle 0, \mathbf{a} \rangle \equiv \top \equiv \langle 0, \mathbf{b} \rangle = emb(\mathbf{b})$ and thus $emb(\mathbf{a}) \wedge emb(\mathbf{b}) \equiv \top$.
  Below we give the truthtables of $\cdot$ and $\rightarrow$.

  We turn to the merger:

$$emb(a \cdot b) = \langle 1 \, _1{\leftarrow} \, a \cdot b, a \cdot b \rangle$$
$$\leq \langle (1 \, _1{\leftarrow} \, b) \, _1{\leftarrow} \, a, a \cdot b \rangle$$
$$= \langle (1 \, _1{\leftarrow} \, a) \wedge ((1 \, _1{\leftarrow} \, b) \leftarrow a), a \cdot b \rangle$$
$$= emb(a) \cdot emb(b).$$

·	0	1	**a**	**b**	⊤
0	0	0	0	0	0
1	0	1	**a**	**b**	⊤
a	0	**a**	**a**	⊤	⊤
b	0	**b**	⊤	**b**	⊤
⊤	0	⊤	⊤	⊤	⊤

→	0	1	**a**	**b**	⊤
0	⊤	⊤	⊤	⊤	⊤
1	0	1	**a**	**b**	⊤
a	0	0	**a**	0	⊤
b	0	0	0	**b**	⊤
⊤	0	0	0	0	⊤

Table 12.3
Truthtables for · and →.

If we have the $OTAT$ property for b we find: $(1 \leftarrow b) \leq 1$ and hence $1 \;_1\!\!\leftarrow a \cdot b = (1 \;_1\!\!\leftarrow b) \leftarrow a$. So in this case: $emb(a \cdot b) = emb(a) \cdot emb(b)$.

EXAMPLE 12.2 Let $\mathcal{Z}$ be the structure $\langle \mathbf{Z}, \mathbf{min}, \mathbf{max}, +, \mathbf{0}, -, -\rangle$, (So $m \to n = n \leftarrow m = n - m$). Note that our ordering is the converse of the usual ordering on the integers. (We use boldface to emphasize that arithmetical objects and operations are intended and not those of the algebra. So e.g. $x + y = \min(x, y)$ and $x \cdot y = x + y$.) It is not difficult to see that $\mathcal{Z}$ is a reduced residuation lattice satisfying Ω. We have:

$$emb(-1) = \langle 1, -1 \rangle,$$
$$emb(0) = \langle 0, 0 \rangle,$$
$$emb(-1 + 1) = \langle 0, 0 \rangle = 1,$$
$$emb(-1) \cdot emb(1) = \langle 1, -1 \rangle \cdot \langle 0, 1 \rangle = \langle \mathbf{max}(1, 0 - -1), 0 \rangle = \langle 1, 0 \rangle.$$

Note that $\langle 0, 0 \rangle < \langle 1, 0 \rangle$.

Clearly $-1 + 1 = 0$, represents a flagrant violation of $OTAT$, since -1 is not a state, but 0 is. More on $\mathcal{Z}$ in section 12.6.4.

Finally suppose A is a (reduced) action lattice. We have:

$emb(a)^* = \langle pre(a), a \rangle^* = \langle (1 \;_1\!\!\leftarrow a) \;_1\!\!\leftarrow a^*, a^* \rangle.$

Now $(1 \;_1\!\!\leftarrow a) \;_1\!\!\leftarrow a^* = (1 \;_1\!\!\leftarrow a^*) \wedge (1 \;_1\!\!\leftarrow a^* \cdot a)$. Moreover $a^* \cdot a \leq a^* \cdot a^* \leq a^*$, so:

$(1 \;_1\!\!\leftarrow a) \;_1\!\!\leftarrow a^* = 1 \;_1\!\!\leftarrow a^* = pre(a^*).$

Ergo $emb(a^*) = emb(a)^*$.

12.4.3   *emb* on States

Let s be a state of $\mathcal{A}$. As is easily seen $emb(s) = \langle 1, s \rangle$. On the other hand if $\langle a', a \rangle \leq \langle 1, 1 \rangle$, then $1 \leq a' \leq 1$ and so $a = a' \cdot a \leq 1$. So $\langle a', a \rangle$ is of the form $\langle 1, s \rangle$. So emb maps the states of $\mathcal{A}$ surjectively on the states of $\mathcal{U}(\mathcal{A})$ minus 0. Obviously emb is injective modulo $\equiv$. We have:

$$emb(s \vee u) = emb(s) \vee emb(u),$$
$$emb(s \wedge u) = \langle 1, s \wedge u \rangle = \langle 1, N_1(s) \wedge N_1(u) \rangle = emb(s) \wedge emb(u),$$
$$emb(s \cdot u) = \langle 1, s \cdot u \rangle = emb(s) \cdot emb(u),$$
$$\langle 1, u \rangle \leftarrow \langle 1, s \rangle = \langle 1, (1 \rightarrow 1) \wedge (1 \rightarrow (1 \cdot u \leftarrow s)) \rangle = \langle 1, u \;_1\!\leftarrow s \rangle$$
$$= emb(u \;_1\!\leftarrow s)$$
$$\langle 1, s \rangle \rightarrow \langle 1, u \rangle = \langle 1 \cdot s, 1 \cdot s \rightarrow 1 \cdot u \rangle = \langle s, s \rightarrow u \rangle.$$

Note that:

$$\langle 1, s \rangle \rightarrow_{\langle 1,1 \rangle} \langle 1, u \rangle = \langle s \vee 1, N_1(1) \wedge N_1(s \rightarrow u) \rangle$$
$$= \langle 1, s \rightarrow_1 u \rangle$$
$$= emb(s \rightarrow_1 u).$$

We may conclude that $\lambda a.(1 \wedge emb(a))$ is an isomorphism between $\mathcal{A}_1$ and $\mathcal{U}(\mathcal{A})_1 \setminus \{0\}$. A trivial consequence of the results of 12.4.3 is:

FACT 12.5 SMP_1 (and hence SMP) is preserved under $\mathcal{U}$.

What happens when we repeat $\mathcal{U}$? Under the right circumstances we get nearly the same algebra. The precise identity is spoiled by the addition of new bottom elements in the construction.

To make our question sensible we must make sure that the circumstances that make $\mathcal{U}$ meaningful and possible are preserved by $\mathcal{U}$. Since we don't know whether Ω is preserved, the most reasonable option is to assume that the principle SMP_1 (which is preserved and implies Ω) holds in $\mathcal{A}$. So assume $\mathcal{A}$ satisfies SMP_1.

Consider the transitions $\mathcal{A} \mapsto \mathcal{U}(\mathcal{A}) \mapsto \mathcal{U}\mathcal{U}(\mathcal{A})$. Let's label the corresponding embeddings by superscript 1 respectively 2.

Let's compute $pre^2(\langle a', a \rangle)$. (Remember that $a' \cdot a \leq 1$, i.e., $a' \leq 1 \leftarrow a$.)

$$\langle 1, 1 \rangle \leftarrow \langle a', a \rangle = \langle 1, (1 \rightarrow a') \wedge (1 \rightarrow (1 \cdot 1 \leftarrow a)) \rangle$$
$$= \langle 1, a' \wedge (1 \leftarrow a) \rangle$$
$$= \langle 1, a' \rangle.$$

Since $\langle 1, a' \rangle$ is already a state, we find: $pre^2(\langle a', a \rangle) = \langle 1, a' \rangle$

Let's assume $\mathcal{U}(\mathcal{A})$ has a new top. Now the elements of $\mathcal{U}\mathcal{U}(\mathcal{A})$ have one of the forms:

- $\langle 0, \top \rangle$ (note that $pre^2(\top) = 0$),

- $\langle \langle 1, s \rangle, \langle a', a \rangle \rangle$ with $\langle 1, s \rangle \leq pre^2(\langle a', a \rangle)$, i.e. $s \leq a'$,

- or $\langle \langle 1, s \rangle, 0 \rangle$, or $\langle 0, \langle a', a \rangle \rangle$, or $\langle 0, 0 \rangle$ or 0.

Trivially for any α, β of the appropriate kind: $\langle 0, \alpha \rangle \equiv \langle 0, \beta \rangle$. So we may ignore the elements of the form $\langle 0, \langle a', a \rangle \rangle$ and $\langle 0, \top \rangle$ in favor of $\langle 0, 0 \rangle$. $\langle \langle 1, s \rangle, 0 \rangle \equiv \langle \langle 1, s \rangle, a \rangle$ precisely if $a = 0$.

FACT 12.6   For $s \leq a'$: $\langle \langle 1, s \rangle, \langle a', a \rangle \rangle \equiv \langle \langle 1, s \rangle, \langle s, a \rangle \rangle$.

Proof $\langle s, a \rangle$ is of the right form, since $s \cdot a \leq a' \cdot a \leq 1$. $\langle \langle 1, s \rangle, \langle s, a \rangle \rangle$ is of the right form since: $\langle 1, s \rangle \cdot \langle s, a \rangle = \langle (s \, _1 \leftarrow s), s \cdot a \rangle = \langle 1, s \cdot a \rangle \leq \langle 1, 1 \rangle$. We check the equivalence: $\langle 1, s \rangle \cdot \langle a', a \rangle = \langle (a' \, _1 \leftarrow s), s \cdot a \rangle = \langle 1, s \cdot a \rangle = \langle 1, s \rangle \cdot \langle s, a \rangle$.   ■

We find:

$$emb^2(\langle s, a \rangle) = \langle pre^2(\langle s, a \rangle), \langle s, a \rangle \rangle = \langle \langle 1, s \rangle, \langle s, a \rangle \rangle.$$

Also:

$$emb^2(0) = \langle pre^2(0), 0 \rangle = \langle \langle 1, 1 \rangle, 0 \rangle.$$

And:

$$emb^2(\top) = \langle pre^2(\top), \top \rangle = \langle 0, \top \rangle \equiv \langle 0, 0 \rangle.$$

So the image of $\mathcal{U}(\mathcal{A})$ covers $\mathcal{U}\mathcal{U}(\mathcal{A})$ except for the elements of the form $\langle \langle 1, s \rangle, 0 \rangle$ for $s \neq 1$.

If $\mathcal{A}$ had a bottom 0, then the top of $\mathcal{U}(\mathcal{A})$ would be $\langle 0, 0 \rangle$ and $emb^2(\langle 0, 0 \rangle) = \langle \langle 1, 0 \rangle, \langle 0, 0 \rangle \rangle$. So in this case $\langle 0, 0 \rangle$ of $\mathcal{U}\mathcal{U}(\mathcal{A})$ would not be in the image of $\mathcal{U}(\mathcal{A})$, since it is above the image $\langle \langle 1, 0 \rangle, \langle 0, 0 \rangle \rangle$ of the top of $\mathcal{U}(\mathcal{A})$.

12.5   Further Properties

We study some specific properties further constraining residuation lattices.

12.5.1 S-injectivity

S-injectivity is the property: $s \cdot a \le s \cdot b \Longrightarrow a \le b$. Note that this is equivalent to: $(s \to s \cdot b) = b$. Ω is a trivial consequence of S-injectivity. Note that for $\langle a', a \rangle$ and $\langle b', b \rangle$ in U: $\langle a', a \rangle \le \langle b', b \rangle \leftrightarrow b' \le a'$ and $a \le b$. Clearly S-injectivity is only a reasonable property if $\mathcal{A}$ has no bottom: the only residuation lattice with bottom satisfying S-injectivity is the degenerated one.

Our primary example of a reduced residuation is the algebra $\mathcal{Z}$ of example 12.2, see further, e.g., section 12.6.4.

How does $\mathcal{U}(\mathcal{A})$ look if $\mathcal{A}$ satisfies S-injectivity?

LEMMA 12.2 $(s \to (s \cdot b \leftarrow a)) = b \leftarrow a$.

Proof

$$
\begin{aligned}
x \le (s \to (s \cdot b \leftarrow a)) &\Longleftrightarrow s \cdot x \le s \cdot b \leftarrow a \\
&\Longleftrightarrow s \cdot x \cdot a \le s \cdot b \\
&\Longleftrightarrow x \cdot a \le b \\
&\Longleftrightarrow x \le b \leftarrow a.
\end{aligned}
$$

∎

As is easily seen using 12.2 our results simplify to:

$$
\begin{aligned}
\langle a', a \rangle \vee \langle b', b \rangle &:= \langle a' \wedge b', a \vee b \rangle \\
\langle a', a \rangle \wedge \langle b', b \rangle &:= \langle a' \vee b', a \wedge b \rangle \\
\langle a', a \rangle \cdot \langle b', b \rangle &:= \langle a' \wedge (b' \leftarrow a), a \cdot b \rangle \\
\langle b', b \rangle \leftarrow \langle a', a \rangle &:= \langle b', (b' \to a') \wedge (b \leftarrow a) \rangle \\
\langle a', a \rangle \to \langle b', b \rangle &:= \langle b' \cdot a, a \to b \rangle \text{ if } b' \le a', \ := 0 \text{ otherwise.}
\end{aligned}
$$

It is immediate that here *emb* commutes also with $\wedge$.

12.5.2 S-idempotency

S-idempotency is the property: $\forall s \in S\, s \cdot s = s$. A primary example of an S-idempotent residuation lattice is any Heyting Algebra $\mathcal{H}$ with $\cdot := \wedge$. Another example is the residuation lattice $\mathcal{REL}$ of binary relations on a given non-empty domain with order $\subseteq$ and as merger relation composition. Yet another example is the algebra of 12.1.

A noteworthy fact is that under certain circumstances $\cdot$ and $\wedge$ will coincide.

FACT 12.1 $s \cdot a \le 1 \Longrightarrow s \cdot a = s \wedge a$.

Proof Clearly:

$$s \cdot a = (s \cdot s) \cdot a = s \cdot (s \cdot a) \leq s \cdot 1 = s \quad \text{and} \quad s \cdot a \leq 1 \cdot a \leq a,$$

so $s \cdot a \leq s \wedge a$. Conversely:

$$s \wedge a \leq (s \wedge a) \cdot (s \wedge a) \leq s \cdot a. \qquad \blacksquare$$

Note that 12.1 tells us that $\mathcal{A}_1$ is in fact a Heyting Algebra! We show that S-idempotency implies SMP (and hence Ω).

THEOREM 12.3 $(a \; {}_1{\leftarrow} \; s) \cdot s = (a \leftarrow s) \wedge s = s \wedge a$ (and similarly for $\rightarrow$).

Proof $(a \; {}_1{\leftarrow} \; s) \cdot s \leq (a \leftarrow s) \cdot s \leq a$ and $(a \; {}_1{\leftarrow} \; s) \cdot s \leq 1 \cdot s = s$, so $(a \; {}_1{\leftarrow} \; s) \cdot s \leq s \wedge a$. Also by our previous lemma: $(a \; {}_1{\leftarrow} \; s) \cdot s = (a \; {}_1{\leftarrow} \; s) \wedge s = (a \leftarrow s) \wedge s \geq a \wedge s.$ $\blacksquare$

We show that under S-idempotency there are very pleasant designated representatives of our $\equiv$-equivalence classes:

THEOREM 12.4 For any $\langle a', b \rangle \in U$, there is a unique $a \leq a'$, such that $\langle a', b \rangle \equiv \langle a', a \rangle$.

Proof Consider any $\langle a', b \rangle \in U$. We have $a' \cdot b = a' \cdot a' \cdot b = a' \cdot (a' \wedge b)$, so $\langle a', b \rangle \equiv \langle a', a' \wedge b \rangle$. Take $a := a' \wedge b$. Clearly $\langle a', a \rangle \in U$. Finally suppose that for some $c \leq a'$ $a' \cdot b = a' \cdot c$. Then $a = (a' \wedge a) = a' \cdot a = a' \cdot c = (a' \wedge c) = c$. $\blacksquare$

For representing the update functions in X, we can replace U by $U^i := \{ \langle a', a \rangle \mid a \leq a' \leq 1 \}$. We will do this in our representation of the operations in $\mathcal{U}(\mathcal{A})$.

A simple calculation shows that the new operations are:

$$\langle a', a \rangle \vee \langle b', b \rangle := \langle a' \wedge b', (a \vee b) \wedge a' \wedge b' \rangle$$
$$\langle a', a \rangle \wedge \langle b', b \rangle := \langle a' \vee b', (a' \vee b') \wedge (a' \rightarrow a) \wedge (b' \rightarrow b) \rangle$$
$$\langle a', a \rangle \cdot \langle b', b \rangle := \langle a' \wedge (b' \leftarrow a), a \wedge b \rangle$$
$$\langle b', b \rangle \leftarrow \langle a', a \rangle := \langle b', b' \wedge (b' \rightarrow a') \wedge (b' \rightarrow (b \leftarrow a)) \rangle$$
$$\qquad\qquad = \langle b', b' \wedge a' \wedge (b \leftarrow a) \rangle$$
$$\langle a', a \rangle \rightarrow \langle b', b \rangle := \langle b' \wedge a, b \wedge a \rangle \text{ if } b' \leq a',$$
$$\qquad\qquad := 0 \text{ otherwise.}$$

Note that $\cdot$ becomes idempotent (and thus S-idempotency is preserved by $\mathcal{U}$). Note also that $\mathcal{U}(\mathcal{A})$ is completely determined by $\mathcal{A}_1$, i.o.w. $\mathcal{U}(\mathcal{A})$ is isomorphic to $\mathcal{U}(\mathcal{A}_1)$. As we noted $\mathcal{A}_1$ is a Heyting Algebra.

12.6   Examples

In this section we introduce some motivating examples.

12.6.1   The Simplest Cases

Consider the trivial one point Heyting Algebra $\mathcal{TRIV}$. It is easily seen that $\mathcal{U}(\mathcal{TRIV})$ is precisely the two point Heyting Algebra $\mathcal{CLASS}$ of Classical Logic, where $+ = \wedge$. Let's look at $\mathcal{U}(\mathcal{CLASS})$. The elements are $\top := \langle 0,0 \rangle$, $1 := \langle 1,1 \rangle$, $\mathbf{a} := \langle 1,0 \rangle$ and 0, where $0 < \mathbf{a} < 1 < \top$. We give the truthtables for $\mathcal{U}(\mathcal{CLASS})$ in table 12.4.

$\cdot$	0	a	1	$\top$
0	0	0	0	0
a	0	a	a	a
1	0	a	1	$\top$
$\top$	0	$\top$	$\top$	$\top$

$\rightarrow$	0	a	1	$\top$
0	$\top$	$\top$	$\top$	$\top$
a	0	$\top$	$\top$	$\top$
1	0	a	1	$\top$
$\top$	0	0	0	$\top$

$\leftarrow$	0	a	1	$\top$
0	$\top$	$\top$	$\top$	$\top$
a	0	1	1	$\top$
1	0	a	1	$\top$
$\top$	0	a	a	$\top$

Table 12.4
Truthtables for $\mathcal{U}(\mathcal{CLASS})$.

Another salient four point algebra has the same elements and the same ordering as $\mathcal{U}(\mathcal{CLASS})$. Moreover the merger is also the same except that $\mathbf{a} \cdot \top = \top$. These data determine the free commutative and idempotent residuation (action) algebra (lattice) on zero generators. (In this business even free algebras on zero generators tend to get fairly complicated. Our example here is the only simple one that I am aware of.)

12.6.2   A DRT-like Semantics for the Positive Part of Predicate Logic

Let *Dref* be a (finite or infinite) set of Discourse Referents (or: Variables). Let a non empty domain D be given. An assignment is a partial function from *Dref* to D. The set of assignments is *Ass*. We write $f \leq g$ for g extends f. A subset F of *Ass* is *persistent* if: $f \in F$ and $f \leq g \Longrightarrow g \in F$.

Let $Ass_V := \{\, f \mid V \subseteq \mathrm{dom}(f) \,\}$. The objects of our algebra are pairs $\sigma = \langle V, F \rangle$, where V is a finite set of discourse referents and where F is a persistent set of partial assignments $f \in Ass_V$. We call these pairs M-states and we call the set of these pairs M. The idea is that V forms a context: it contains the mental objects that are present. The set F is a constraint on these objects, which codifies the actual information.

We do not demand: $f \in F \Longrightarrow f \mid V \in F$ (*Downwards Projection Property* or *DPP*). The reason is that a collection of M-states with this extra property would (with the obvious ordering; see below) not form a reduced residuation lattice. Absence of the *DPP* means that some variables are constrained even if they are not present in V. One could say that these variables are virtually present.

Define:

$$\langle V', F' \rangle \leq \langle V, F \rangle :\Longleftrightarrow V \subseteq V' \text{ and } F' \subseteq F.$$

Define:

$$\langle V, F \rangle \wedge \langle V', F' \rangle = \langle V \cup V', F \cap F' \rangle,$$
$$\langle V, F \rangle \vee \langle V', F' \rangle = \langle V \cap V', F \cup F' \rangle.$$

It is easy to see that $\wedge$ and $\vee$ are indeed inf and sup for our ordering. (Note that $\wedge$ preserves the *DPP*, but $\vee$ does not.)

We take $\cdot := \wedge$. We see that we may take $1 := \langle \emptyset, Ass \rangle$. Note that $1 = \top$ w.r.t. our ordering, so M is really a set of states in our sense.

We show that the resulting structure is a reduced Heyting Algebra by giving $(\sigma \rightarrow \tau)$. Let F be any set of assignments, V any finite set of Discourse Referents. Define: $Int_V(F) := \{ f \in Ass_V \mid \forall g \geq f \ g \in F \}$.

Suppose $\sigma = \langle V, F \rangle$ and $\tau = \langle W, G \rangle$. We show:

$$(\sigma \rightarrow \tau) = \langle W \setminus V, Int_{W \setminus V}(F^c \cup G) \rangle.$$

Consider $\rho = \langle X, H \rangle$. We find:

$$\rho \wedge \sigma \leq \tau \Longleftrightarrow W \subseteq X \cup V \text{ and } H \cap F \subseteq G$$
$$\Longleftrightarrow W \setminus V \subseteq X \text{ and } H \subseteq (F^c \cup G).$$

Note that if $W \setminus V \subseteq X$, then $H = Int_X(H) = Int_{W \setminus V}(H)$; also note that $Int_{W \setminus V}$ is monotonic (w.r.t. $\subseteq$) and idempotent. We find:

$$\rho \wedge \sigma \leq \tau \Longleftrightarrow W \setminus V \subseteq X \text{ and } H \subseteq Int_{W \setminus V}(F^c \cup G)$$
$$\Longleftrightarrow \rho \leq \langle W \setminus V, Int_{W \setminus V}(F^c \cup G) \rangle.$$

Note that $Int_{W \setminus V}(F^c \cup G) = \{ h \in Ass_{W \setminus V} \mid \forall f \geq h \ (f \in F \Longrightarrow f \in G) \}$.

Thus we consider the reduced residuation lattice:

$$\mathcal{M}\text{-}\mathcal{S} := \langle M\text{-states}, \vee, \wedge, \wedge, 1, \rightarrow, \rightarrow \rangle.$$

12.6.2.1 Special Elements Given a set W of Discourse Referents, we define $\bot :=$ $\langle \emptyset, \emptyset \rangle$. Note that we use $\bot$ in a non-standard way, since it is not the bottom w.r.t. our ordering! $\top_\sigma := \bot \to \sigma$, $\bot_\sigma := \bot \wedge \sigma$.

It is easily seen that if $\sigma = \langle V, F \rangle$, then $\top_\sigma = \langle V, Ass_V \rangle$, $\bot_\sigma = \langle V, \emptyset \rangle$, $\bot = \bot_\bot$. Note that $\top = 1 = \top_1$. We have e.g.:

- $\bot_\sigma \leq \sigma \leq \top_\sigma$,
- $\top_\sigma \wedge \top_\tau = \top_{\sigma \wedge \tau}$,
- $(\bot_\sigma \to \tau) = (\bot_\sigma \to \bot_\tau) = (\sigma \to \top_\tau) = (\top_\sigma \to \top_\tau) = \top_{\sigma \to \tau}$.

We write $\sim \sigma$ for $(\sigma \to \bot)$.

FACT 12.2 σ has the *DPP* iff $\top_\sigma \leq (\sigma \vee \sim\sigma)$

Proof Left to the reader.   ∎

We turn to $\mathcal{U}(\mathcal{M}\text{-}\mathcal{S})$.

EXCURSION 12.6.2.2 (ON THE CONNECTION WITH DRT) The elements of U are our semantic counterparts of DRS's. The relationship with semantic counterparts à la Zeevat is as follows. The objects we get as interpretations of predicate logical formulas (if an appropriate dynamical implication is added) have the form $\langle \top_\sigma, \tau \rangle$, i.o.w. $\langle \langle V, Ass_V \rangle, \langle W, G \rangle \rangle$, where $V \subseteq W$. These specific objects can also be written: $\langle V, G, W \rangle$. The DRS meanings in Zeevat's sense are in fact objects of the form $\langle W \setminus V, G \rangle$. So Zeevat just represents the discourse referents that are newly introduced ($\approx$ bound and active variables), but not the active variables that are imported ($\approx$ free variables). This difference leads to a slightly different logic.

12.6.2.3 Sample Meanings Let a suitable a first order model with domain D be given. We represent meanings as certain update functions on M-states, viz. as elements of $\mathcal{U}(\mathcal{M}\text{-}\mathcal{S})$. We exhibit some representations of meanings. We confuse a variable v with $\langle \{v\}, \emptyset \rangle$. So $v = \bot_v$.

$$\|\exists v\| := \langle \top, \top_v \rangle \qquad (\text{``a'' as in ``a man''}),$$
$$\|\iota v\| := \langle \top_v, \top_v \rangle \qquad (\text{``he/she/it'', and ``the''}),$$
$$\|P(v,w)\| := \langle \top_{v \wedge w}, \langle \{v, w\}, \{\, f \in Ass_{\{v,w\}} \mid \langle f(v), f(w) \rangle \in \|P\| \,\} \rangle \rangle$$
$$\qquad\qquad (\text{where } \|P\| \text{ is some binary relation on } D \text{ associated to } P$$
$$\qquad\qquad \text{in the given model}),$$
$$\|\varphi.\psi\| := \|\varphi\| \cdot \|\psi\|.$$

We compute the meaning of $\exists v.P(v,w)$. Let $F := \{\, f \in Ass_{\{v,w\}} \mid \langle f(v), f(w)\rangle \in \|P\|\,\}$.

$$\begin{aligned}
\|\exists v.P(v,w)\| &= \langle \top, \top_v\rangle \cdot \langle \top_{v\wedge w}, \langle\{v,w\}, F\rangle\rangle \\
&= \langle \top \wedge (\top_v \to \top_{v\wedge w}), \top_v \wedge \langle\{v,w\}, F\rangle\rangle \\
&= \langle \top_w, \langle\{v,w\}, F\rangle\rangle.
\end{aligned}$$

This means that the meaning of $\exists v.P(v,w)$ presupposes that w has a value and 'produces' a value for v.

12.6.2.4 Restriction Can we define $\Psi_\alpha \mid \mathrm{dom}(\Psi_\beta)$ in our framework? Yes. Let $\alpha = \langle \sigma', \sigma\rangle$ and let $\beta = \langle \tau', \tau\rangle$. $\Psi_\alpha \mid \mathrm{dom}(\Psi_\beta) = \Psi_{\langle \sigma' \wedge \tau', \sigma \wedge \tau'\rangle}$. Note that:

$$\langle \tau', \tau\rangle \leftarrow \langle \tau', \tau\rangle = \langle \tau', \tau' \wedge \tau' \wedge (\tau \leftarrow \tau)\rangle = \langle \tau', \tau'\rangle.$$

Also:

$$\langle \tau', \tau'\rangle \cdot \langle \sigma', \sigma\rangle = \langle \tau' \wedge (\sigma' \leftarrow \tau'), \tau' \wedge \sigma\rangle = \langle \sigma' \wedge \tau', \sigma \wedge \tau'\rangle.$$

So the pair representing $\Psi_\alpha \mid \mathrm{dom}(\Psi_\beta)$ is $(\beta \leftarrow \beta) \cdot \alpha$. (Note that we only used S-idempotency.)

12.6.2.5 Conditions The function D with $\sigma D := \bot_\sigma$ can be viewed as producing the domain of σ. An update function F on M-states is a condition if it doesn't change the domains of its inputs, i.o.w. $F \circ D = D \mid \mathrm{dom}(F)$. Clearly D itself is a condition, since it is idempotent and everywhere defined.

In $\mathcal{U}(\mathcal{M}\text{-}\mathcal{S})$, D can be represented by $\delta := \langle \top, \bot\rangle$. So α is a condition if:

$$\alpha \cdot \delta = (\alpha \leftarrow \alpha) \cdot \delta.$$

So $\langle \sigma', \sigma\rangle$ represents a condition if $\langle \sigma', \sigma\rangle \cdot \langle \top, \bot\rangle = \langle \sigma', \sigma'\rangle \cdot \langle \top, \bot\rangle$, i.e.

$$\langle \sigma' \wedge (\sigma \to \top), \sigma \wedge \bot\rangle = \langle \sigma', \sigma \wedge \bot\rangle = \langle \sigma', \sigma' \wedge \bot\rangle = \langle \sigma' \wedge (\sigma' \to \top), \sigma' \wedge \bot\rangle,$$

in other words $\sigma D = \sigma' D$.

Note that in our sample meanings above $\|\exists v\|$ is a state and $\|\iota v\|$ and $\|P(v,w)\|$ are conditions.

12.6.2.6 Discussion

i. I don't think the present approach is completely satisfactory. For one thing it has the consequence that if a variable v that is already active is existentially quantified, then the quantifier is ignored. E.g. $\|P(v).\exists v.Q(v)\| = \|P(v).Q(v)\|$. Thus the present treatment gives the existential quantifier the meaning: *introduce as new if not already present, otherwise ignore.* A second objection is the presence of the ill understood virtual referents due to the absence in general of the *DPP*. Still I think it is worthwile to pursue the present approach a bit, since it is the simplest approach known to existential quantification, that gets the presuppositional aspect right. For another aproach see e.g. Vermeulen [17] or my forthcoming *Meanings in Time*.

ii.   We do not go into the treatement of validity and dynamic implication here (see also section 12.2.5.1). As far as I can see to do this reasonably one should extend the algebraic framework. E.g. van Benthem suggests to define validity using the precondition operator $\diamond$. $\diamond\langle\sigma',\sigma\rangle$ gives the weakest precondition for getting truth (rather than definedness) after applying $\Psi_{\langle\sigma',\sigma\rangle}$. We have $\sigma \leq \diamond\langle\sigma',\sigma\rangle \leq \sigma'$. We can proceed to define validity in the style of Groenendijk & Stokhof as:

$$\alpha \models_\sigma \beta :\Longleftrightarrow \sigma\Psi_\alpha \leq \diamond\Psi_\beta.$$

12.6.3   The Algebra of Relations $\mathcal{REL}$

Let $\mathcal{REL}_D := \langle \wp(D \times D), \cup, \cap, \circ, ID, \rightarrow, \leftarrow \rangle$, where D is some non-empty set. $\wp(D \times D)$ is the set of binary relations on D and ID is the identity relation or i.o.w. the diagonal. The residuations are given by:

$$u(R \rightarrow S)v :\Longleftrightarrow \forall w\,(wRu \Longrightarrow wSv)$$
$$u(S \leftarrow R)v :\Longleftrightarrow \forall w\,(vRw \Longrightarrow uSw).$$

In this structure the states are precisely the tests: subsets of the diagonal. Relation Composition is idempotent on the states. Thus $\mathcal{U}(\mathcal{REL}_D) = \mathcal{U}(\mathcal{SET}_D)$, where $\mathcal{SET}_D$ is the residuation lattice $\langle \wp(D), \cup, \cap, \cap, D, \rightarrow, \rightarrow \rangle$, and $X \rightarrow Y := X^c \cup Y$.

Since the states are tests, I think the ordering on $\mathcal{REL}_D$ is wrong for doing Dynamic Semantics. It is an open problem to find an ordering on (a suitable subset of) $\wp(D \times D)$ that does what we want.

12.6.4   Simple Stacking Cells

The monoid of Simple Stacking Cells (*SSC*'s) is simply the free monoid on two generators (and), satisfying the equation () = 1. (In this monoid we follow the usual convention of notationally suppressing $\cdot$.) *SSC*'s are in a sense the integers of the well-known bracket test.

An application of *SSC*'s is to be found in my forthcoming paper *Meanings in Time*.

We will show that *SSC*'s can be viewed as integers under a presupposition. Before proceeding to the formal development, let me first sketch the basic intuition.

Consider a clerk in charge of a store of items. At certain times either a number of items is demanded or a number of items is delivered. Whenever more items are asked for than are in store, the clerk's firm will go bankrupt. We assume that delivery may not be postponed. Now consider a sequence of demands and deliveries. The effect of such a sequence can be well described by two numbers: first the minimal number that has to be in store for the factory not to go bankrupt in the process. Secondly the sum of the deliveries minus the sum of the demands. The two numbers, say m and n, together give us an update function for numbers s of items in store: if $m \leq s$ then the result of updating is $s + n$, the result is undefined otherwise. Note that we use 'undefined' to model bankruptcy.

If we equate *more in store* with *more informed*, then it becomes clear that the order on the integers needed to model the clerk example is the converse of the usual ordering. This leads us to consider the reduced residuation lattice $\mathcal{Z} = \langle \mathbf{Z}, \min, \max, +, \mathbf{0}, -, - \rangle$ as our initial structure. The states of $\mathcal{Z}$ are the $s \leq 1$, i.e. the $s \geq 0$, i.o.w. precisely the non-negative integers. $\mathcal{U}(\mathcal{Z})$ is precisely the appropriate structure to model the process of deliveries an demands.

Note that if we would allow the clerk to postpone giving out items that are asked for untill sufficiently many items are in store the appropriate objects to describe the process would be simply the integers.

To return to our original way of describing the SSC's every string of brackets can be rewritten using the conversion [() $\mapsto$ empty string] to a string)$\ldots$)($\ldots$(of first m right brackets and then k left brackets. The corresponding representative in $\mathcal{U}(\mathcal{Z})$ will be $\langle m, k - m \rangle$.

Since $\mathcal{Z}$ satisfies S-injectivity, the operations of $\mathcal{U}(\mathcal{Z})$ take the particularly simple form given in section 12.5.1.

12.6.5   A Clerk with Several Kinds of Items

What happens if the clerk of example 12.6.4 watches over a store of different kinds of items, say apples, pears, bananas, etc.? The obvious idea is to model the store as a *multiset* of kinds. A multiset can be represented as a function from the possible kinds to the natural numbers $\mathbf{N}$. We could allow that there is an infinity of possible kinds. Since we are only interested in finite multisets, we stipulate that the representing functions are almost everywhere $\mathbf{0}$. We are thus lead to the following idea.

Let K be the (possibly) infinite set of the possible kinds that may be in store. For any reduced residuation lattice we define $\mathcal{FINSUP}(K, \mathcal{A})$ to be the reduced residuation lattice of the functions of finite support from K to $\mathcal{A}$. A function has finite support if its value is 1 almost everywhere. The operations of $\mathcal{FINSUP}(K, \mathcal{A})$ are the pointwise induced ones. Note that if K is infinite and if $\mathcal{A}$ is non-trivial, then $\mathcal{FINSUP}(K, \mathcal{A})$ has no bottom. It is easily seen that this construction preserves properties like S-injectivity, SMP, Ω, etc.

We now model our clerk's work by $\mathcal{B} := \mathcal{U}(\mathcal{FINSUP}(K, \mathcal{Z}))$. As soon as of any kind more is asked than is in store the firm will go bankrupt. An alternative modelling is:

$C := \mathcal{FINSUP}(K, \mathcal{U}(\mathcal{Z})) = \mathcal{FINSUP}(K, \mathcal{SSC})$. We leave it to the reader to trace the similarities and differences between these solutions.

As a specific un-clerklike example of our clerk, consider the strictly positive rational numbers. The inverse divisibility ordering on the strictly positive rationals $\mathbf{Q}^+$ is defined as follows:

$$m/n \le i/j :\Longleftrightarrow i \times n \text{ divides } m \times j \qquad (m, n, i, j \in \mathbf{N} \setminus \{0\}).$$

Note that if we assume that m and n, resp. i and j have no common divisors this simplifies to:

$$m/n \le i/j :\Longleftrightarrow i \text{ divides } m \text{ and } n \text{ divides } j.$$

Let P be the set of primes. Consider $\mathcal{FINSUP}(P, \mathcal{Z})$; this is, say, $\langle Q, \vee, \wedge, \cdot, 1, \rightarrow, \leftarrow \rangle$. It is easy to see that $\langle \mathbf{Q}^+, \le, \times \rangle$ is isomorphic to $\langle Q, \le, \cdot \rangle$ (where the $\le$ of the last structure is derived from $\vee$). So $\langle \mathbf{Q}^+, \le, \times \rangle$ can be extended (uniquely) to a reduced residuation lattice. Implication is division here. $\mathcal{U}(\mathcal{FINSUP}(P, \mathcal{Z}))$ becomes a kind of presuppositional version of the non-negative rationals. Note that the 0 of this last structure is like the usual **0** of the rationals.

QUESTION 12.2   Can one make a presuppositional version of all the rationals with both operations $+$ and $\times$, which has a 'reasonable' structure?

12.7   An Alternative Representation of the Update Functions

In this section we try to increase the analogy of our update functions to the integers.

Again fix a reduced residuation algebra $\mathcal{A}$ satisfying Ω.

It could be felt as a defect that we didn't really give a construction of actions from states, but of actions from actions. Thus it fails the analogy with the construction of the integers from the natural numbers. An alternative representation of the update functions goes some way to remedy this defect, but not in all cases the full way .

Consider $\Psi_{\langle a', A \rangle}$ for $\langle a', A \rangle \in U$. Note that (in the presence of Ω) $\Psi_{\langle a', A \rangle}$ is fully detemined by a' and $a' \cdot A$. Thus we can represent our update functions also by pairs of states $[a', a]$, where $a' \le 1$ and for some A: $a = a' \cdot A \le 1$. Let V be the set of these pairs. Below we will describe $\mathcal{U}(\mathcal{A})$ in terms of these alternative pairs on the assumption of SMP.

We transform our earlier results to the new format:

$$[a', a] \le [b', b] \Longleftrightarrow b' \le a' \text{ and } b' \cdot A \le b$$
$$\Longleftrightarrow b' \le a' \text{ and } (b' {}_1\!\leftarrow a') \cdot a \le b$$
$$[a', a] \vee [b', b] = [a' \wedge b', (a' \wedge b') \cdot (A \vee B)]$$

$$
\begin{aligned}
&= [a' \wedge b', ((a' \wedge b')\ {}_1\!\leftarrow a') \cdot a \vee ((a' \wedge b')\ {}_1\!\leftarrow b') \cdot b)] \\
&= [a' \wedge b', (b'\ {}_1\!\leftarrow a') \cdot a \vee (a'\ {}_1\!\leftarrow b') \cdot b)] \\
[a',a] \wedge [b',b] = {}&[a' \vee b', (a' \vee b') \cdot ((a' \to a) \wedge (b' \to b))] \\
[a',a] \cdot [b',b] = {}&[a' \wedge (b' \leftarrow A), (a' \wedge (b' \leftarrow A)) \cdot A \cdot B] \\
&= [((b' \leftarrow A)\ {}_1\!\leftarrow a') \cdot a', ((b' \leftarrow A)\ {}_1\!\leftarrow a') \cdot a' \cdot A \cdot B] \\
&= [(b'\ {}_1\!\leftarrow a) \cdot a', (b'\ {}_1\!\leftarrow a) \cdot a \cdot B] \\
&= [(b'\ {}_1\!\leftarrow a) \cdot a', (a\ {}_1\!\leftarrow b') \cdot b' \cdot B] \\
&= [(b'\ {}_1\!\leftarrow a) \cdot a', (a\ {}_1\!\leftarrow b') \cdot b] \\
[b',b] \leftarrow [a',a] = {}&[b', b' \cdot ((b' \to a') \wedge (b' \to (b \leftarrow A)))] \\
&= [b', b' \cdot (b' \to (a' \wedge (b \leftarrow A)))] \\
&= [b', b' \cdot (b' \to ((b \leftarrow A)\ {}_1\!\leftarrow a') \cdot a')] \\
&= [b', b' \cdot (b' \to ((b\ {}_1\!\leftarrow a) \cdot a')] \\
[a',a] \to [b',b] = {}&[b' \cdot A, b' \cdot A \cdot (b' \cdot A \to b)] \\
&= [(b'\ {}_1\!\leftarrow a') \cdot a, (b'\ {}_1\!\leftarrow a') \cdot a \cdot ((b'\ {}_1\!\leftarrow a') \cdot a \to b)] \\
&\quad \text{if } b' \le a', \\
&= 0 \text{ otherwise.}
\end{aligned}
$$

So we find:

$$
\begin{aligned}
[a',a] \le [b',b] :&\Longleftrightarrow b' \le a' \text{ and } (b'\ {}_1\!\leftarrow a') \cdot a \le b \\
[a',a] \vee [b',b] :&= [a' \wedge b', (b'\ {}_1\!\leftarrow a') \cdot a \vee (a'\ {}_1\!\leftarrow b') \cdot b] \\
[a',a] \wedge [b',b] :&= [a' \vee b', (a' \vee b') \cdot ((a' \to a) \wedge (b' \to b))] \\
[a',a] \cdot [b',b] :&= [(b'\ {}_1\!\leftarrow a) \cdot a', (a\ {}_1\!\leftarrow b') \cdot b] \\
[b',b] \leftarrow [a',a] :&= [b', b' \cdot (b' \to (b\ {}_1\!\leftarrow a) \cdot a')] \\
[a',a] \to [b',b] :&= [(b'\ {}_1\!\leftarrow a') \cdot a, (b'\ {}_1\!\leftarrow a') \cdot a \cdot ((b'\ {}_1\!\leftarrow a') \cdot a \to b)] \\
&\quad \text{if } b' \le a', \\
:&= 0 \text{ otherwise.}
\end{aligned}
$$

Note that in case $\mathcal{A}$ satisfies S-idempotency, our new representation collapses to our earlier one since $a = a' \cdot A = a' \wedge A$.

The new way of representing is not really a construction from actions from states, since (i) V is still generally dependent on all of $\mathcal{A}$ and (ii) in our formulations unrelativized pre-implications still occur in an essential way. In case $\mathcal{A}_1$ is idempotent a simple inspection shows that the dependency disappears. This leads us to the following list of problems.

12.7.1   Open Problems

i.   Under what conditions can the dependence of $\mathcal{U}(\mathcal{A})$ on the full pre-implications of $\mathcal{A}$ be eliminated?

ii.   What $V_\mathcal{B}$ are possible? Are there nice properties characterizing possible $V_\mathcal{B}$?

iii.   What is the structure of the $V_\mathcal{B}$ with $\subseteq$?

It is easy to see that $V_\mathcal{B}$ is isomorphic to $V_{\mathcal{U}(\mathcal{B})}$.

12.7.2   The Alternative Representation of $\mathcal{SSC}$

It is easy to see that $V_\mathcal{Z}$ is $\{\, [a', a] \mid a', a \in \mathbf{N} \,\}$. A simple computation gives simplified relations and operations on $\mathcal{SSC}$ $(= \mathcal{U}(\mathcal{Z}))$:

$$[a', a] \leq [b', b] \Longleftrightarrow b' \leq a' \text{ and } (b' \,_1\!\leftarrow a') \cdot a \leq b$$
$$[a', a] \vee [b', b] = [a' \wedge b', (b' \,_1\!\leftarrow a') \cdot a \vee (a' \,_1\!\leftarrow b') \cdot b)]$$
$$[a', a] \wedge [b', b] = [a' \vee b', (a' \vee b') \cdot ((a' \rightarrow a) \wedge (b' \rightarrow b))]$$
$$[a', a] \cdot [b', b] = [(b' \,_1\!\leftarrow a) \cdot a', (a \,_1\!\leftarrow b') \cdot b]$$
$$[b', b] \leftarrow [a', a] = [b', (b \,_1\!\leftarrow a) \cdot a']$$
$$[a', a] \rightarrow [b', b] = [(b' \,_1\!\leftarrow a') \cdot a, b] \text{ if } b' \leq a', = 0 \text{ otherwise.}$$

These operations on the representing pairs are fully in terms of $\mathcal{Z}_1$, i.e. the reduced residuation algebra $\mathcal{N} := \langle \mathbf{N}, \mathbf{min}, \mathbf{max}, +, \mathbf{0}, \dot{-}, \dot{-} \rangle$ (where $\dot{-}$ is cut-off substraction). So $\mathcal{SSC}$ satisfies the ideal of being constructible in a way analoguous to the construction of $\mathbf{Z}$. More information on $\mathcal{SSC}$ in the appendix 12.8.

12.7.3   Negative Strings?

A rather obvious idea is to go and use our framework to create *negative strings*. However it turns out that the relevant analogue of $\mathcal{N}$ is not a reduced residuation algebra, but just a model of process algebra, since pre-implication is lacking. Thus this problem escapes our present framework. I hope to report on this puzzle in a later publication.

12.8   Appendix: Subalgebras of $\mathcal{SSC}$

In this appendix we collect some data on the $\mathcal{SSC}$'s.

It is instructive to write out the representation of section 12.7 of SSC's in purely numerical terms. Thus an SSC α is either 0 or $\top$ or a pair $[\mathsf{pop}_\alpha, \mathsf{push}_\alpha]$. We specify the relations and operations on the pairs.

$$\alpha \leq \beta \Longleftrightarrow \mathsf{pop}_\alpha \leq \mathsf{pop}_\beta \text{ and } \mathsf{push}_\beta \leq \mathsf{push}_\alpha + (\mathsf{pop}_\beta \doteq \mathsf{pop}_\alpha)$$

$$\alpha \vee \beta = [\mathbf{max}(\mathsf{pop}_\alpha, \mathsf{pop}_\beta),$$
$$\mathbf{min}(\mathsf{push}_\alpha + (\mathsf{pop}_\beta \doteq \mathsf{pop}_\alpha), \mathsf{push}_\beta + (\mathsf{pop}_\alpha \doteq \mathsf{pop}_\beta))]$$

$$\alpha \wedge \beta = [\mathbf{min}(\mathsf{pop}_\alpha, \mathsf{pop}_\beta),$$
$$\mathbf{min}(\mathsf{pop}_\alpha, \mathsf{pop}_\beta) + \mathbf{max}(\mathsf{push}_\alpha - \mathsf{pop}_\alpha, \mathsf{push}_\beta - \mathsf{pop}_\beta)]$$

$$\alpha \cdot \beta = [\mathsf{pop}_\alpha + (\mathsf{pop}_\beta \doteq \mathsf{push}_\alpha), \mathsf{push}_\beta + (\mathsf{push}_\alpha \doteq \mathsf{pop}_\beta)]$$

$$\beta \leftarrow \alpha = [\mathsf{pop}_\beta, \mathsf{pop}_\alpha + (\mathsf{push}_\beta \doteq \mathsf{push}_\alpha)]$$

$$\alpha \rightarrow \beta = [\mathsf{push}_\alpha + (\mathsf{pop}_\beta \doteq \mathsf{pop}_\alpha), \mathsf{push}_\beta] \text{ if } \mathsf{pop}_\alpha \leq \mathsf{pop}_\beta,$$
$$= 0 \text{ otherwise.}$$

Some well-known algebras are subalgebras of SSC for all operations, with the exception of one of the implications.

The algebra $\mathcal{POP}$ is given by 0, $\top$ and the pairs of the form $[\mathsf{pop}, 0]$. A simple computation shows:

$$\alpha \leq \beta \Longleftrightarrow \mathsf{pop}_\alpha \leq \mathsf{pop}_\beta$$

$$\alpha \vee \beta = [\mathbf{max}(\mathsf{pop}_\alpha, \mathsf{pop}_\beta), 0]$$

$$\alpha \wedge \beta = [\mathbf{min}(\mathsf{pop}_\alpha, \mathsf{pop}_\beta), 0]$$

$$\alpha \cdot \beta = [\mathsf{pop}_\alpha + \mathsf{pop}_\beta, 0]$$

$$\beta \leftarrow \alpha = [\mathsf{pop}_\beta, \mathsf{pop}_\alpha]$$

$$\alpha \rightarrow \beta = [\mathsf{pop}_\beta - \mathsf{pop}_\alpha, 0] \text{ if } \mathsf{pop}_\alpha \leq \mathsf{pop}_\beta, = 0 \text{ otherwise.}$$

The true internal post-implication of $\mathcal{POP}$ is the maximum in $\mathcal{POP}$ below $\beta \leftarrow \alpha$ $(= [\mathsf{pop}_\beta, \mathsf{pop}_\alpha])$. We have:

$$[p, 0] \leq [\mathsf{pop}_\beta, \mathsf{pop}_\alpha] \leftrightarrow p + \mathsf{pop}_\alpha \leq \mathsf{pop}_\beta.$$

So $\alpha \rightarrow \beta$ is the maximum of the elements of $\mathcal{POP}$ below $\beta \leftarrow \alpha$. (The treatment of 0 and $\top$ is as is to be expected.) The resulting algebra is the $+, \max$ residuation lattice. If we rename $0 =: \perp$ and identify $[n, 0]$ with n, the ordering of this algebra looks as follows: $\perp, 0, 1, 2, \ldots, \top$. $\cdot$ is addition, where $\perp + \alpha = \alpha + \perp = \perp$ and for $\alpha \neq \perp$: $\top + \alpha = \alpha + \top = \top$. Finally both residuations are given by: $\top \doteq \alpha = \top$, $\alpha \doteq \perp = \top$, $\alpha \doteq \beta = \perp$ if $\alpha < \beta$, and $m \doteq n = m - n$ if $n \leq m$. Good alternative notations for $\perp$ and $\top$ here would have been $-\infty$ and ∞.

Next consider the algebra $\mathcal{PUSH}$ consisting of the elements 0 and $[0, p]$. We have:

$$\alpha \leq \beta \iff \mathsf{push}_\beta \leq \mathsf{push}_\alpha$$
$$\alpha \vee \beta = [0, \mathbf{min}(\mathsf{push}_\alpha, \mathsf{push}_\beta)]$$
$$\alpha \wedge \beta = [0, \mathbf{max}(\mathsf{push}_\alpha, \mathsf{push}_\beta)]$$
$$\alpha \cdot \beta = [0, \mathsf{push}_\beta + \mathsf{push}_\alpha]$$
$$\beta \leftarrow \alpha = [0, \mathsf{push}_\beta \mathbin{\dot{-}} \mathsf{push}_\alpha]$$
$$\alpha \rightarrow \beta = [\mathsf{push}_\alpha, \mathsf{push}_\beta].$$

Clearly $\mathcal{PUSH}$ is just $\mathcal{SSC}_1$ (with the proper internal pre-implication, viz. $\rightarrow_1$ and the proper treatement of 0). The resulting algebra is the $+$, min residuation lattice or the tropical residuation lattice. If we identify $[0, n]$ with n, this looks as follows. The ordering is $\bot, \ldots, \mathbf{2}, \mathbf{1}, \mathbf{0}$. $\cdot$ is addition. The residuations are both cut off substraction $\dot{-}$ on the natural numbers and $\alpha \mathbin{\dot{-}} \bot = \mathbf{0}$, $\bot \mathbin{\dot{-}} n = \bot$. (An alternative notation for $\bot$ could have been ∞.)

Finally consider $\mathcal{ZERO}$ the subalgebra given by 0, $\top$ and the pairs $[z, z]$ (or $\langle z, \mathbf{0} \rangle$). We have:

$$\alpha \leq \beta \iff z_\alpha \leq z_\beta$$
$$\alpha \vee \beta = [\mathbf{max}(z_\alpha, z_\beta), \mathbf{max}(z_\alpha, z_\beta)] = \langle \mathbf{max}(z_\alpha, z_\beta), \mathbf{0} \rangle$$
$$\alpha \wedge \beta = [\mathbf{min}(z_\alpha, z_\beta), \mathbf{min}(z_\alpha, z_\beta)] = \langle \mathbf{min}(z_\alpha, z_\beta), \mathbf{0} \rangle$$
$$\alpha \cdot \beta = [\mathbf{max}(z_\alpha, z_\beta), \mathbf{max}(z_\alpha, z_\beta)] = \langle \mathbf{max}(z_\alpha, z_\beta), \mathbf{0} \rangle$$
$$\beta \leftarrow \alpha = [z_\beta, \mathbf{max}(z_\alpha, z_\beta)] = \langle z_\beta, z_\alpha \mathbin{\dot{-}} z_\beta \rangle$$
$$\alpha \rightarrow \beta = [z_\beta, z_\beta] = \langle z_\beta, \mathbf{0} \rangle \text{ if } z_\alpha \leq z_\beta, \ = 0 \text{ otherwise.}$$

Note that $[z, z] \leq [u, v] \leftrightarrow z \leq u$ and $v \leq u$. So $\alpha \rightarrow \beta$ is the maximal element in $\mathcal{ZERO}$ below $\beta \leftarrow \alpha$. So identifying $[z, z]$ with z the ordering of our algebra is: $\bot, \mathbf{0}, \mathbf{1}$, $\ldots, \top$. $\cdot$ is $\mathbf{max}$ on $\mathbf{N} \cup \{\top\}$, but $\bot \cdot \alpha = \alpha \cdot \bot = \bot$. Finally both residuations are equal and are as described by $\chi(\alpha \leq \beta) \cdot \beta$, where $\chi(\alpha \leq \beta) := \mathbf{0}$ if $\alpha \leq \beta$, $:= \bot$ otherwise.

Acknowledgements

I thank Kees Vermeulen, Jan van Eijck and Johan van Benthem for stimulating discussions. (Both readings of the previous phrase are correct.) I thank Marcus Kracht for his incisive comments, often critical, always inspiring. I thank Vaughan Pratt for introducing me to the concept of Action Algebra.

References

[1] Abrusci, V. M., 1991, *Phase Semantics and Sequent Calculus for Pure Noncommutative Classical Linear Propositional Logic*, JSL 56, 1403–1451.

[2] Benthem, J. F. A. K. van, 1991, *Language in action*, North Holland, Amsterdam.

[3] Bergstra, J. A. & Tucker, J. V., 1990, *The inescapable stack, an exercise in algebraic specification with total functions*, Report P8804b, University of Amsterdam.

[4] Conway, J. H., 1971, *Regular Algebra and Finite Machines*, Chapman and Hall, London.

[5] Eijck, J. van, editor, 1990, *Proceedings on Logics in AI, JELIA '90*, Springer Lecture Notes in Computer Science 478.

[6] Eijck, J. van & Vries, F. J. de, 1992, *Dynamic Interpretation and Hoare Deduction*, JoLLI, I, 1–44.

[7] Fernando, T., 1992, *Transition systems and dynamic semantics*, Logic and AI, D. Pearce and G. Wagner, eds. (LNCS 633), Springer, Berlin, pp. 232-251.

[8] Frege, G., 1975, *Funktion, Begriff, Bedeutung*, edited by G. Patzig, Vandenhoeck & Ruprecht, Göttingen.

[9] Fitting, M., 199?, *Kleene's Logic, Generalized* (forthcoming).

[10] Groenendijk, J. & Stokhof, M., 1991, *Dynamic Predicate Logic*, Linguistics and Philosophy 14, 39–100.

[11] Kozen, D., 1981, *A Completeness Theorem for Kleene Algebras and the Algebra of Regular Events*, Proceedings of the 6th IEEE Symposium on Logic in Computer Science, 214–225.

[12] Kozen, D., 1992, *On Action Algebras*, these proceedings.

[13] Leeuwen, J. van, editor, 1990, *Formal Models and Semantics, Handbook of Theoretical Computer Science, volume B*, Elsevier, Amsterdam.

[14] Perrin, D., 1990, *Finite Automata*, in Van Leeuwen [13, 1–57].

[15] Pratt, V., 1990, *Action Logic and Pure Induction*, in Van Eijck [5, 97–120].

[16] Vermeulen, C., 1991a, *Sequence Semantics for Dynamic Predicate Logic*, Logic Group Preprint Series 60, Department of Philosophy, University of Utrecht.

[17] Vermeulen, C., 1991b, *Merging without Mystery*, Logic Group Preprint Series 70, Department of Philosophy, University of Utrecht.

[18] Visser, A., 199?, *Meanings in Time*, forthcoming.

[19] Zeevat, H. W., 1991, *A compositional version of Discourse Representation Theory*, Linguistics and Philosophy 12, 95–131.

[20] Zeinstra, L., 1990, *Reasoning as Discourse*, Master's Thesis, Department of Philosophy, University of Utrecht.

.

DATE DUE